Higher Education and the Growth of Knowledge

This book sketches the history of higher education in parallel with the development of science. Its goal is to draw attention to the historical tensions between the aims of higher education and those of science in the hope of contributing to improving the contemporary university. A helpful tool in analyzing these intellectual and social tensions is Karl Popper's philosophy of science demarcating science and its social context. Popper defines a society that encourages criticism as "open," and argues convincingly that an open society is the most appropriate one for the growth of science. A "closed society," on the other hand, is a tribal and dogmatic society. Despite being the universal home of science today, the university, as an institution that is thousands of years old, carries traces of different past cultural, social, and educational traditions. The book argues that, by and large, the university was, and still is, a closed society and does not serve the best interests of the development of science and of students' education.

Michael Segre is Professor of the History of Science at the "Gabriele D'Annunzio" University in Chieti. He has taught for many years at the University of Munich, and has been fellow of the Harvard University Center for Italian Renaissance Studies, Florence. He is the author of numerous books and articles, including *In the Wake of Galileo* (1991) and *Peano's Axioms in their Historical Context* (1994).

Routledge Studies in Cultural History

For a full list of titles in this series, please visit www.routledge.com

Higher Education and the Growth of Knowledge

A Historical Outline of Aims and Tensions

Michael Segre

Routledge
Taylor & Francis Group

LONDON AND NEW YORK

First published 2015 by Routledge

2 Park Square, Milton Park, Abingdon, Oxfordshire OX14 4RN
711 Third Avenue, New York, NY 10017

*Routledge is an imprint of the Taylor & Francis Group,
an informa business*

First issued in paperback 2017

Library of Congress Cataloging-in-Publication Data

Segre, Michael, 1950–
    Higher education and the growth of knowledge : a historical outline of aims
and tensions / by Michael Segre.
        pages cm. — (Routledge studies in cultural history ; 38)
    Includes bibliographical references and index.
    1. Education, Higher—Philosophy. 2. Education, Higher—History.
I. Title.
    LB2322.2.S4317 2015
    378.001—dc23
    2015007301

ISBN: 978-0-415-73566-7 (hbk)
ISBN: 978-0-8153-8142-6 (pbk)

Typeset in Sabon
by Apex CoVantage, LLC

MIX
Paper from
responsible sources
FSC
www.fsc.org   FSC™ C013985
Printed in the United Kingdom
by Henry Ling Limited

To the memory of my parents

Contents

Contents

Introduction

It is with mixed feeling that I undertake writing this book; it criticizes an institution, the university, to which I have belonged for more than four decades and toward which I owe a profound debt. Let me here emphasize my respect for the university as one of the basic institutions of our civilization. Moreover, Karl Popper, whose philosophy inspired this book, regards criticism an expression of esteem, and my criticism is intended as a contribution to improve the university, and not to belittle its profound achievements in any way.

Admittedly, beyond love for learning, from the beginning of my university studies and, throughout my academic carrier, I have felt uneasiness. As a student, I often found lectures boring or incomprehensible, teachers authoritarian and not collaborative, and textbooks or scholarly texts unreadable. I detested, above all, exams, marks, and pressure in general. My students, 40 years later, express the same general complaint. As an academic, too, I have an uneasy feeling. I wonder why an institution intended to contribute to science and to the intellectual and personal growth of its members should produce so much (needless) social competition, servility, and stress.

The philosophy of Karl Popper, which I began understanding in depth only at an advanced stage of my career thanks to the teaching of Joseph Agassi (who had been, among other, Popper's assistant), helped me decipher some of my uneasiness. Popper's masterpiece *The Open Society and its Enemies* (hereafter *OS*), written during WWII and published in 1945, distinguishes between the "closed society," which is tribal and dogmatic, and the "open society," which is the society that encourages criticism. Popper argues convincingly that the open society is more appropriate for the development of science and democracy. He nevertheless admits that there is no totally closed or open society; there are, rather, some societies more open than others. Popper's social theory of knowledge opened my eyes to see that, perhaps, the university is still not sufficiently open enough to fulfill its duty as the universal institution monopolizing, as it does, science.

Popper's social philosophy is part and parcel of his philosophy of science, formulated initially in his *Logik der Forschung* (1934, translated into English as *The Logic of Scientific Discovery*). This last fundamental work argues against the hoary view that scientific discovery is observation laden. According to Popper, theories are not constructed inductively from single observations, as still believed on the popular level—just as observing a large number of white swans does not ascertain that all swans are white. Instead of beginning a research with observation, Popper suggests the following heuristic: formulate a question to nature; guess an answer that can form a theory. Empiricism (observation or experiment) plays a role in testing the theory, and a theory, according to Popper, is scientific only if testable, with clear conditions of refutation. Refutations delimit the applicability of a theory, or reject it and call for a better one. This is how science evolves, possibly endlessly. Consequently, Popper argues against the common view that science is the knowledge of truth: it is only the pursuit of truth, ever less wrong, coming closer and closer, but quite likely never reaching it. Yet it is the duty of the scientist to persist in this quest. This philosophy, as Ian Jarvie has shown in his masterly *The Republic of Science. The Emergence of Popper's Social View of Science 1935–1945* (2001), is inseparable from its social context, and science can grow better if it can be freely subjected to criticism, for example, in an open society. When speaking of science in this book, I refer to science according to Popper's view of science.

What appealed to me in Popper's philosophy of science is his fallibilism: science, just like man, and common sense, is fallible. Instead of seeking and demanding perfection in learning (which cannot be reached and serves entirely as an excuse for the imposition of authority), as often universities request, Popper more humbly suggests to improve what is wrong, and this, I feel, should be the way universities should act and improve themselves.

To be sure, Popper's philosophy has been debated and criticized. Indeed Popper's views are not faultless, though they are the best available and generally accepted, albeit grudgingly. One of the criticisms to Popper's philosophy of science is, for example, that just like it is hopeless to ascertain that a theory is true, one cannot be certain it can be disproved, although Popper says explicitly that only contradictions are fully refutable (e.g., all swans are white and this swan is not white) and offers a rule for accepting refutations, and in general, a convention in science (Popper 1995, § 6, §11, §31–46). Yet Popper is snubbed above all for his being an iconoclast. He points out, on many occasions, how "the emperor is naked."

My work as a historian made me realize that much of the "closure" of universities belongs to the past and is practiced as no more than a very bad habit. Being aware of this could help improving matters. This book is a historical outline attempting to show how many closed aspects of the university comprise a superfluous heritage. I restrict myself to the Western higher learning since University is mainly a Western creation, though there have been higher learning institutions in other civilizations, such as China. Moreover, I restrict myself further to a few aspects of past Western higher learning with some examples. I did not, and could not, delve into the detailed study of primary sources: the history of Western higher learning is vast, and such a study would have required much more than a lifetime. I highlight, rather, some of the well-known history. The historical aspects I wish to emphasize are mainly related to the historical monopoly of the clergy over higher learning with the consequent dogmatic and unscientific approach, the restrictive guild-like form of the university, and the confusion between "liberal" and vocational learning. This alone is enough to render the university a relatively closed institution, not entirely appropriate to the development of science and the intellectual and personal growth of students.

Chapter 1 is introductory. It points out how some basic concepts related to higher learning, such as "science" or "institution," are not easy to define, and how basic aspects of higher education are rooted in the dawn of civilization.

The next two chapters deal with the classical roots, still very present. The ancient Greeks introduced the liberal arts, which are non-vocational (today's universities pretend, often deceitfully, to be vocational) but gave rise to modern science. Chapter 2 deals, following Popper, with the roots of the open and closed societies in Athens at the time of Socrates and Plato, the latter, according to Popper, being the first theorist of the closed society. We do not know how exactly Plato's celebrated Academy functioned, but he introduced, on the theoretical level at least, deliberate stress insinuated into studies, and thereby endeavored to form a meritocracy. A meritocracy normally aims at conforming to a framework, whereas the modern university and science demand and deserve excellence—which breaks away from frames. Chapter 3 deals with Aristotle, the founder of the Lyceum, whose philosophy is most relevant to the medieval university and, more generally, to Western civilization, and cannot be neglected. Aristotle, like Plato, has been subject to many interpretation and misrepresentations, and the Aristotle I present has no claim to be historically genuine. Rather the Aristotle I present is closer to the traditional, historically debatable presentation, whose legend, however, I find more relevant to the dawn of the university. With Aristotle observation may have become the main source of knowledge. Aristotle's tendency to classify fields, is also still very present, and eventually discourages interdisciplinarity, which is one of the most fertile driving forces of science, as it leaves no room thereof, except, possibly in philosophy proper.

The next two chapters deal with the religious roots of higher education in early Jewish and Islamic higher learning. It is amazing how the ancient Jewish learning is neglected in this context despite that it is generally agreed that the Jews are one of the most literate people in history. The Jews may have been, inter alia, the first to introduce the title in higher learning, the *semikha*, that is, the ordaining of rabbis as scholars and public teachers of the law, rather than as priests. Jewish higher learning, to which chapter 4 is devoted, concentrated on (religious) law studies; and so did Islamic higher learning (chapter 5). There is no direct evidence that the Jewish higher learning influenced the Islamic one and that both influenced medieval universities. Yet there have undoubtedly been contacts and mutual influences. The University of Bologna, considered the first university in history, was initially a school of higher law studies. More in general, religion, guild-like association, and degree granting are general features of medieval universities (Chapter 6) that left traces till the present day. To be sure, the first universities were corporative, vocational, though not necessarily religious, institutions, but were soon taken over by the clergy. They also could not avoid introducing the non-vocational, but essential, teaching of the liberal arts. And modern science emerged out of the liberal arts, although outside universities. Yet the confusion between vocational and "liberal" fields is still present and disregarded by contemporary universities.

A clerical guild such as the university was not suitable to foster the emerging Renaissance knowledge. More flexible institutions, such as courts, learned societies, or artisan's workshops were more appropriate springboards (chapters 7–10). Private learned societies, for example, were less subject to associative constraints—whether clerical or lay—, career needs, or dogmatic thought, whether religious or profane. Many contributions to higher learning came from purely vocational schools (chapter 8), which were traditionally undervalued and remained such till the 19th century, though they were the basis for modern higher engineering schools. Technology and science require a dialogue that goes beyond traditional barriers and clichés (Chapters 9 and 10), and modern science was generated primarily outside the university, and sometimes despite the university. Galileo, who can be considered the first modern scientist, was a champion in overcoming intellectual, social, and professional barriers. Yet science, as a social enterprise, requires meeting centers, and some scientific institutions that undertook that role were chartered and became major carriers of modern science: the best known are the Royal Society of London and the French Académie des Sciences (Chapter 10). But their official recognition inevitably introduced social boundaries that might have been among the causes of their decline. In a period in which the university as a center of higher learning declined, vocational French engineering schools, either military or civil, founded mainly by the French old regime, began to flourish and were further developed and fostered by the Revolution and the Napoleonic empire. Their success was among the driving forces behind the establishment of the modern "Humboldtian" university (Chapter 11). Wilhelm von Humboldt realized the need of a free environment for the growth of knowledge and strove to apply it to University. This brought the universities' rebirth, though some conservative, traditional elements remained and are still present.

This book does not produce an agenda for the improvement of the modern university, except a call to open up to discussion about possible reforms and to pay more attention to the above elements. For my part I feel University, as an institution, can hardly be reformed, and indeed reforms imposed by the state in several countries did not prove very successful. Single universities, whether a state or a private, have their own peculiarity, traditional or innovative, and may, of course, progress. What is of basic importance is that a university should clearly define and explain its purposes and the way to achieve them.

This book is based on my Italian book *L'università aperta e i suoi nemici*, published in 2013, and I am indebted to the Carabba publisher for having given me permission to use parts of it. Gail McDowell translated the text into English, and I elaborated the translation and adapted it to a more general scholarly and possibly popular level. Both books were inspired by Joseph Agassi, who read the typescript and commented on it. Max Novick from Routledge and Tina Cottone from Apex encouraged the project and were always helpful. Alan Musgrave and Bill Shea suggested improvements, and Maurice Finocchiaro made invaluable remarks. Giulio Lucchetta put at my disposal his knowledge of classical philosophy. Sonja Brentjes, Emmanuel Sivan, and Paola Pizzo suggested improvements to Chapter 5 on Islam. Aaron Agassi edited my typescript. My thanks to them all.

I am indebted to the University of Chieti for having granted me a sabbatical during the academic year 2012–2013 and for having funded several travels overseas for the purpose of my research. The Duke University invited me to take advantage of its outstanding libraries. The National Library of Turin and the library of humanities at the University of Chieti provided me with material quickly and efficiently.

In quoting material not originally in English I have used the available translations. Otherwise, translations are mine, and I quoted the original source in relevant occasions. My knowledge of Hebrew allowed me to express Hebrew words in Latin alphabet as spoken in modern Hebrew, without transliterating. Not knowing Arabic, however, I quoted the transliteration I found in the literature. I have also reported, in brackets, the dates of birth and death of most of the people mentioned (of reign in case of monarchs or rulers) who are no longer alive. In many cases these dates are not known with precision, and "c." for "circa" may refer to birth as well as to death.

Appendix 1, concerning Galileo's current historiography, is an English version of my article "Galilei und die Medici: Post-Renaissance Mäzenatentum oder Postmoderne Geschichtsschreibung?" published in Gudrun Wolfschmidt (ed.), *Astronomisches Mäzenatentum* (2008); Appendix 2 is based on a lecture I held in 2012 at the Max Planck Institute for the History of Science in Berlin at the workshop "Towards a history of the history of science: 50 years since 'Structure.'" It will be published with the proceedings, and I am grateful to the Max Planck Institute for having given me permission for advance republication within this book.

1   Then and Now

The aims and context of higher education are not entirely clear

The university, an institution created in the Middle Ages that today produces, teaches, and dominates science worldwide, is complex and fraught with contradiction. How has this come to pass?

To begin with, there is still no general consensus as to what exactly science is (or should be) and what are its boundaries, beyond which non-scientific knowledge and learning begin. Moreover, the university at times teaches and cultivates knowledge that is clearly nonscientific, such as theology. I will not enter here into a discussion concerning past or present opinions about science. Let me, for the moment, refer to science in as broad a sense as possible, and in a way that perhaps is better described by the German term *Wissenschaft*, including not only the "hard" natural or exact sciences (in German, *Naturwissenschaften*), but also other branches of knowledge such as the "soft" social sciences and the "softer" humanities.

Science as a process remains an aspect of culture. And culture cannot subsist without a society that creates, practices, sustains, and disseminates said culture. Likewise, no society can exist without a culture. Every society, in order to maintain its stability, or even only to survive, must create institutions for the transmission of culture from one generation to the next. It may seem that at least in the very early stages of humanity, knowledge sustained itself: children learned what their parents knew by imitating them and joining them in their daily activities. And yet, this is not so: even from the earliest stages of human development, the instituted role of the guardian of knowledge existed. The Old Testament, for instance, forbade some specific types of knowledge, and the priests (*cohanim*, plural of *cohen*) were a paradigm of guardians of knowledge. In the Greek world, the Pythagorean sect jealously kept its discoveries secret. And Aristotle's pupil Aristoxenus writes in the 4th century BC that Plato, a century earlier, wanted to burn all the works of the pre-Socratic Democritus. Perhaps this was because Plato was an idealist, and Democritus was a materialist (however, the story is a matter of controversy and may be apocryphal). But we can safely put all of this aside, because our concern here is with a literate society that is much more developed: we tacitly take it for granted, in what follows, that we are speaking only of advanced literate societies. Any society, be it an entire civilization or just a close-knit group, is inevitably structured by conflict, and its members are likely to advance different values or experience different needs. And this is mirrored in the tension of conflict structured into their educational institutions. Hence the challenge presented by the very topic of this book.

What, then, are the mechanisms of the growth, evolution, and transmission of knowledge? What is, and what should be, the interaction between educational institutions and the societies to which they belong? What knowledge should be transmitted and how? What should the relationship be between the inculcation of values and the cultivation of vocational competences? What form should elementary or advanced educational institutions assume? These and other questions concern the university; the answers are still far from satisfactory.

These questions are joined by more fundamental issues; besides the question related to the nature of science, which will be treated later in this book, a basic sociological question is, what is an institution? Is it a social structure such as a school or a bank? Historically, the question about them was, can an institution have an aim or a mission? What, for example, is the mission of universities? Are institutions formed on the basis of need, or do they grow naturally? Does an institution create opportunities, and thus takes on the proverbial life all its own? Does an institution need to be chartered for specific optimal performance? The answers may vary depending on the theories or social doctrines. Holistic ("holism" derives from the Greek expression *to holon*, which means "the whole") philosophers and sociologists such as Karl Marx (1818–1883) or Émile Durkheim (1858–1917), for instance, maintain that the behavior of individuals is determined by society, even as though society had a soul. Therefore, even institutions such as universities, could have a conscience, that is, a mission. On the other hand, individualist philosophers maintain that only individuals, and not society, have the faculty to decide, and thus, an institution cannot have a mission of its own but may only follow the mission, or the missions, that are individually set by individuals, especially its own members. Proponents of this latter group include John Stuart Mill (1806–1873), the eclectic Max Weber (1864–1920), and, more recently, the defender of liberalism Friedrich A. von Hayek (1899–1992), winner of the Nobel prize for Economics, and the philosopher Karl R. Popper (1902–1994). Mill even maintains that institutions do not actually exist. Naturally, there is also a middle ground, but the debate is open, and, therefore, to hold that universities might have a mission of their own could prove to be a misleading starting point. For the sake of simplicity I will adopt for *institution* the simple and cautious definition suggested by the philosopher Joseph Agassi, rejecting the holistic view of them as super-individuals and the individualist view of them as sub-individuals, he views them as inter-individual means of encounter between individuals.

Today's higher learning is historically rooted in a profoundly different and bygone milieu

The institution that we call "University" has roots that reach back to the dawn of time; officially founded in the Middle Ages for vocational purposes, right from the start it absorbed more than a few characteristics of less-utilitarian institutions that had preceded it. This book is an attempt to shed light upon the deep historical roots of the university as an institution, along with their philosophical and social aspects, a source of some of the ambiguities shared today by modern universities. The final objective is to produce some hints toward a project of university study that is able to respond in a more rational way to the intellectual challenges posed in these early years of the third millennium. The first chapters trace the institution's most primordial roots. Those who are well-acquainted with the subject will find here little or nothing that is new, while historians who are specialized in the various fields under examination will most likely find many inaccuracies. It is difficult, if not impossible, to offer a precise and exhaustive presentation of such ancient processes in a few short chapters. Further on in the book, I delve into facets such as the history of the university system, which might prove helpful in explaining some of its modern characteristics, without neglecting the manner in which the social, intellectual, professional, and historical context of the past was profoundly different from today's.

Nowadays, in the Western world, with elementary education imposed by law, illiteracy is virtually non-existent, and in some countries students are required to attend school until they reach eighteen years of age. Modern civilization has also developed a multitude of fields and professions that not only require a long and complex period of learning but also mental faculties, such as abstraction, which might have been unimaginable in the past, even though other faculties doubtless existed back then that have since been lost. In Antiquity, as in the Middle Ages, the priorities of education were far different from those of today; culture was transmitted orally, and literacy was the domain of very few. It is not even certain that literacy was indispensable for activities and occupations that required a high level of expertise. Most likely, many sovereigns in Antiquity were illiterate, even though this did not prevent them from being excellent administrators. Nevertheless, these and other assertions are merely conjectural; the available documentation regarding the ancient world is thin. In any case, it would be inappropriate to overlook the distortions risked for any study of ancient or medieval education, including higher education, from the modern point of view. This is not to deny, of course, that past traditions still influence our ways of thinking and learning.

Ancient civilizations, such as the ones that flourished in the Far East, India, Mesopotamia, Egypt, or other parts of the world, nonetheless, achieved high intellectual results in fields such as literature, mathematics, engineering, architecture, and astronomy. Other fields, such as ancient medicine and the practice of religious cults, required, in their specific contexts, profound knowledge, including knowledge of esoteric and magical rituals. The cultures that performed them must have had institutions that were able to create, develop, and transmit this knowledge, even though we do not know for sure how they functioned. We do not know much about the institutions involved, but these fields form the prehistory of modern higher learning. All the evidence seems nevertheless to point out that possible higher learning was confined to closed religious castes.

When dealing with literate societies, let us recall that the inhabitants of Mesopotamia were the first humans to write, although only a small percentage could do so, and one ancient profession associated with considerable schooling was that of the scribe. Scribes were a clerical caste and did much more than just write: historian of ancient religion Karel van der Toorn (2007, 57) calls them "the academics of their time." They exercised functions analogous to modern civil servants, judges, lawyers, or bankers, forming an elite responsible for administration and accountancy. Literacy and knowledge was a mark of social standing, and kings boasted of their scholarship.

Writing in ancient Mesopotamia meant engraving cuneiform letters on tablets of clay, and not everybody who could read could also write. Learning to write was an enterprise. There were specific schools, known as "tablet houses" (*é-dubba*), which could have been in private houses—where fathers instructed their sons—, or attached to temples—where teachers or older students schooled younger ones. Learning was based on memorizing and went beyond reading and writing: the syllabus included subjects such as grammar, law, administration, mathematics, science, music, and languages. Of course only very few attended them.

Scribes in Mesopotamia would easily find administrative employments. A few continued their studies and specialized in professions such as cult singers, astrologers, diviners, exorcists, or physicians. These higher-level studies included learning specific texts and, at a more advanced level, discussing their meaning. Having completed their studies, scholars could find employment in a temple, although employment was not guaranteed. Within the temple compound they would be associated with a workshop or a library. The first organized library in the world may have been that built by the Assyrian king Ashurbanipal (r. 668–627 BC) in Nineveh (northern Iraq): it contained more than twenty thousand clay tablets (Al-Khalili 2010, 70).

In ancient Egypt, too, the skill of writing was handed down from father—normally a priest who kept books in a temple—to son. There is also evidence of the existence of schools granting elementary education consisting, much like in Mesopotamia, of studying subjects such as geography, arithmetic, and geometry; a group would sit about the teacher and learn by chanting. The next level of studies included memorizing wisdom texts by chanting them. Only later, in order to become scribes, would they be made literate in hieratic, a quick cursive writing, on papyrus by means of reed brush and ink, or on anything that could be used as a writing surface, allowing the performance of administrative tasks. Hieratic developed alongside hieroglyphs, the script used for monumental inscription. To master hieroglyphs required a long schooling of specialization in a high-level profession, such as physician or lector-priest. The study took place in religious centers of scholarship called "Houses of Life," a scriptorium with a library, gathering scholars such as physicians, astronomers, mathematicians, and sculptors who acted as teachers.

Besides writing, one other domain that required much knowledge was the art of healing, precursory to one of the first subjects taught in the university: medicine. In ancient time it, too, was in most cases related to religion, although this varied according to locale, and involved several types of specialists, such as priests, magicians, practical healers, and others. In the old testament, healing was a role assigned to the priests, and they were expected more or less to perform practices that today we would denote as "shamanic." In the Greek world there were sanctuaries dedicated to Asclepius, the god of medicine and healing, where ill and injured people came to seek healing.

Healing and writing were, of course, not the only activities that required high proficiency. In general we do not know exactly how the expertise was acquired. Yet, already at this early stage of civilization, one can single out a few points relevant to the question of the role of the modern university. First, higher learning, from ancient time, has been the domain of clerics and therefore strictly interconnected with religion, a link persisting in universities at least up until the 19th century. Clerics were a caste, the type of society that Popper defines as "closed," that is, dogmatic. Secondly, as early as in ancient time, studies and knowledge were related to social rank, risking the misuse of knowledge for social purposes as remains frequent in modern time. Third, higher education in the ancient world before the appearance of Greek philosophy, being the basis of most modern sciences, seems to have been primarily, but not exclusively, vocational: it granted a profession that enabled one to find employment and also what one could call "wisdom." Things, we shall soon see, started changing in ancient Greece, where the last aspect was emphasized and the foundations of science were laid. Today the university, despite its image as the antechamber of employment, monopolizes the sciences, and not all sciences are necessarily vocational. In an age of analphabetism, literacy was a precious and demanded capacity, just as university training was demanded and guaranteed employment in the 19th century, when Western countries grew quickly and their population was partly illiterate. Today, in the nearly totally literate Western world, university graduates are much less required and often unemployed. Even nowadays, overqualified graduates, in particular, are often therefore rejected by employers; indeed, this was at times already true for Mesopotamian specialized scholars. Fourth, the method of studying, at least in Mesopotamia and Egypt (and most probably in the rest of the ancient world), consisted mainly in memorization, for the culture was primarily an oral one. Memorization in all of its monotony and obsolescence nowadays, has left significant traces in modern higher education.

Things began to change in ancient Greece. The heterogeneous and vibrant classical culture contributed to modern higher education in many ways, some of which are different and even contradictory to the Mesopotamian or Egyptian heritage. It left the traces of institutions of higher learning, which often developed around great thinkers. The basic education of ancient Greece included gymnastics and music (the term *gymnos* means naked: athletes used to train naked; the term *music* derives from the muses, the divinities of the arts). The first of these two disciplines was a function of preparedness for war; the second was much more than what we consider music today, consisting in learning the literary tradition handed down largely in song and dance and in similar rituals. And yet, between the 7th and the 6th centuries BC, in the pre-Socratic era, schools already existed, not so much as teaching institutions, but as groups of promoters of philosophical thought whose influence extended down through generations. These thinkers posed fundamental questions that came to be treated later in time by science.

Relatively little is known about the pre-Socratic philosophers. Most sources are indirect and fragmentary. Thales, who is widely considered the first scientist, is said to be the father of the Ionian tradition, which developed in Miletus, in Asia Minor: Ionia was the region of central coastal Anatolia. With Anaximenes and Anaximander, he attempted to find an answer to the problem of the *arche* (literally "beginning," to denote principal or foremost); in other words, he tried to answer the question, what is the principle or the basic material of reality that lies behind what our senses can perceive?— hardly a vocational activity. On a more institutional level, however, it is not clear what the relationship was between these philosophers, nor how they transmitted their knowledge to their contemporaries.

The same questions were posed by another famous school of philosophy of that same era, founded by the legendary Pythagoras in the 6th century BC in Croton, in southern Italy. The sources narrating its history are uncertain. They tell that the Pythagoreans were a confraternity of men and women, with initiation rites and the communion of goods.

Even though the Pythagoreans already numbered women in their group, until the 20th century only very few women participated in higher education. Their exclusion is a corollary to their discrimination in almost every civilization, which confined them to the semi-reclusion of household chores and modest and marginal activities, often on a par with slavery. Like it or not, higher education was usually reserved exclusively to males. In the early 1900s, women at university were still an exception.

The Pythagoreans believed in the transmigration of the soul; they ate no meat or fava beans, and one of the reasons could have been to avoid favism, a hereditary disorder typical for the neighborhood of Croton appearing with the consumption of fava beans (for the many other possible reasons, see Sole 2004). Above all, they attributed a mystical value to numbers. They investigated the nature of numbers, thereby making important contributions to mathematics. They have been attributed with the "Pythagorean theorem," which remains, inter alia, fundamental to physics: it states that in a right triangle, the square of the hypotenuse is equal to the sum of the squares of the two legs.

The people of Mesopotamia and the Egyptians were already familiar with this theorem but, as far as we know, not with any proof thereof. Nor is it clear whether the Pythagoreans had a concept of proof, but they did develop ideas like the Pythagorean triads (three magnitudes a, b, and c, such that $a^2+b^2 = c^2$, e.g., 3, 4, and 5) and similar series. The most important contribution of the Pythagoreans to civilization is probably that of having ascertained the importance of mathematics. For example, they investigated the relationship between the length of the strings of musical instruments and musical notes, and they gave harmony both a numerical and a universalistic dimension.

The documentation concerning the pre-Socratic philosophers is too thin to tell whether some of them were priests. Certainly, Pythagoras seems to have learnt much from priests in several places. The Neoplatonic philosopher Porphyry (c. 233–305 AD) recounts that in Egypt Pythagoras "lived with the priests, and learned the language and wisdom of the Egyptians, and three kinds of letters, the epistolic, the hieroglyphic, and symbolic, whereof one imitates the common way of speaking, while the others express the sense by allegory and parable."[1] This is strengthened by the following impressive description taken from the long biography of Pythagoras written by the other Neoplatonic philosopher, Iamblichus (c. 250–325 AD):

> He frequented all the Egyptian temples with the greatest diligence and with accurate investigation, he was both admired and loved by the priests and prophets with whom he associated. And having learnt with the greatest solicitude every particular, he did not neglect to hear of any transaction that was celebrated in his own time, or of any man famous for his wisdom, or any mystery in whatever manner it might be performed; nor did he omit to visit any place in which he thought something more excellent might be found. On this account he went to all the priests, by whom he was furnished with the wisdom which each possessed. He spent therefore two and twenty years in Egypt, in the adyta of temples, astronomizing and geometrizing, and was initiated, not in a superficial or casual manner, in all the mysteries of the Gods, till at length being taken captive by the soldiers of Cambyses, he was brought to Babylon. Here he gladly associated with the Magi, was instructed by them in their venerable knowledge, and learnt from them the most perfect worship of the Gods. Through their assistance likewise, he arrived at the summit of arithmetic, music, and other disciplines; and after associating with them twelve years, he returned to Samos about the fifty-sixth year of his age.
>
> (Iamblichus 1818, 9)

(The adyta was a sacred place allowed to priests only; Cambyses II was son and successor to Cyrus, king of Persia). If the above biographies, written many centuries after the pre-Socratic age, were reliable, then the Pythagorean fraternity certainly lived and acted very much like a caste of clerics. Otherwise they might at least point out the importance of the castes of priests in the 3rd century AD. Other pre-Socratic philosophers, and other later Greek philosophers, however, may have been an exception in the norm that intellectuals were clerics.

Some of the questions that the pre-Socratics posed are still current. Parmenides of Elea (c. 515–460 BC; Elea is also located in southern Italy), for example, was concerned with the nature of change. Zeno of Elea (c. 490–430 BC) posed paradoxes, which have become famous, regarding the continuity of motion. Leucippus of Miletus (c. 490–? BC), of whom we know very little, developed the theory of atoms, which was later elaborated by Democritus (c. 470–370 BC). Empedocles (c. 490–430 BC) of Agrigentum (Sicily), however, originated the prevalent theory of the four elements: water, earth, air, and fire. Socrates's and Pericles's contemporary, Hippocrates of Kos (c. 460–370 BC), founded the school of medicine that took his name and is considered the father of Western rational medicine. These and other philosophers might have founded philosophical or medical schools, but we do not know how they transferred their knowledge to their disciples.

Greek learning endeavored to develop virtues

Even though we are not quite familiar with the mechanisms through which the pre-Socratics transferred their knowledge, we do know that itinerant teachers, the Sophists ("those who make a profession of their knowledge"— the Greek terms *sophos* and *sophia* are commonly translated as "wise man" and "wisdom"; *sophizomai* means to "become wise"), established themselves in 5th century BC Greece. They offered to cultivate the virtues, charging their pupils, although providing an education that went beyond the purely elementary. Werner Jaeger (1888–1961), the well-known classicist of the past century, concluded in an in-depth study of Greek literary and philosophical works that even though the Sophist education comprised various means and methods of teaching, it had two fundamentally distinct methods of learning: "One was to import to it an encyclopaedic variety of *facts*, the material of knowledge, and the other to give it *formal* training of various types" (Jaeger 1946–47, 1, 292, his Italics). In modern terms, we could call these two educational methods the imparting erudition and training, respectively.

This brings us to one additional fundamental point: learning can follow at least two primary paths. Along the first the teacher, who could be substituted by a textbook, or even by a machine, transmits knowledge to the student unidirectionally. Basically, the pupil remains a passive vessel, the figurative tabula rasa. In the second, along the path that is formative, there is a more or less complex interaction between teacher and pupil, who must necessarily adopt a more active role, whereas the teacher must be the more nondirective. This path presumes that the pupil has, to begin with, some manner of innate knowledge or comprehension.

To this day, despite the formidable educational means available, there is no valid replacement for the teacher. Yet, like it or not, the university and, more generally, the modern school, is primarily an institution that imparts knowledge rather than trains. It is, without a doubt, cheaper and easier to give students a textbook, have them memorize it and evaluate their expertise (their "professionalism") on the basis of their familiarity with the contents. The teaching process can be perfected with the help of a DVD or a simulator, or with any other means that is able to transmit knowledge. However, just like science itself, the material transmitted unidirectionally, be it by a teacher or by any technological substitute, is never absolute wisdom or truth. As leading mathematician and philosopher Alfred North Whitehead (1861–1947) remarks, "A merely well-informed man is the most useless bore on God's earth" (Whitehead 1967, 1). Actually, according to Popper, the practice of science is nothing other than an infinite chain of trial and error, with no actual guarantee of ever at all approaching truth thereby. Thus, we can hardly train a professional, a physician, a lawyer, or a pilot only through written texts or unidirectional lessons, no matter how good these are.[2] Unidirectional university teaching can reach absurd outcomes, such as those described by Nobel Laureate in physics Richard Feynman (1918–1988) in his autobiography (Feynman 1985, 211–218): when he was teaching in Brazil, he noticed that physics students would learn the stuff by heart and pass exams by reciting it to the teacher, but were unable to solve related elementary problems. Subjects like physics and mathematics, or philosophy, history, and archeology, require an exchange of points of view, and the ancient Greeks already knew this. A professional should be an expert, and an expert must know how to evaluate the context in which he or she operates and be able to evaluate the risks; this is why it is also necessary to have experience that is accumulated through trial and error. Besides knowledge, what is also needed is learning that is *formative,* and this book reflects the hope that this aspect will be furthered, in the university as in other schools. Such a teaching was practiced, among others, by Albert Einstein (1879–1955), but it was an exception, and Einstein's biographer, the positivist physicist Philipp Frank (1884–1966; see Frank 1953, 91) criticizes it as "uneven." The modern university does not only emphasize the acquisition of knowledge; it often confuses plain acquisition of details with actual learning.

To return to the ancient world, the Sophists, as Plato mentions in his work dedicated to one of them, *Protagoras,* were more inclined toward formative learning, indeed the training of the free man (*Protagoras* 312 *b*). Classic literature mentions various teaching methods of the Sophists. One of central importance was perhaps invented by Zeno: dialectics (from the Greek, the art of arguing). There were various types of dialectics, just as, more generally, there are many types of intellectual interaction between individuals. The type Zeno propounded was a *reductio ad absurdum*—deductive reasoning that ends in the absurd: one starts with a hypothesis and deduces an absurd result that refutes the hypothesis. A common example that can be found in modern, secondary school geometry is indirect demonstration, in which one supposes that a statement is false and from this supposition arrives at a contradiction. Protagoras, instead, is supposed to have introduced eristic dialogue (from the Greek *eristiké [téchne]*—"argumentative art"), in which the only goal of the dispute is to prevail over the adversary with one's reasoning.[3]

The Sophists also developed the notion—noticed by the Mesopotamian kings and probably much less obvious back then than it is today—that, besides ensuring social stability, knowledge ennobles; they endow education with a delicate dimension that, over the course of the centuries, was destined to become important and problematic.

The Sophists were not popular with the establishment, and not only because of their greedy art of polemics; by encouraging individual development, they questioned tradition, in particular those traditions that were convenient to certain groups detaining power. This is a first socio-philosophical contradiction at the heart of education as we know it, and of higher education in particular: the purpose of education is to maintain social stability, or rather, to safeguard certain norms; but at the same time, it develops faculties that go beyond these very norms and risk destabilizing society. This contradiction opposes tradition, which gives a sense of security but risks leading to immobility and decadence, and progress, which, in turn, can develop in unexpected and dangerous directions. It also reflects a second unresolved socio-philosophical question, already touched upon, to wit: the dispute between individualism and holism. Is it the individual who determines the form of society (as the individualists believe) or vice versa (as the holists believe)? Can one find a balance between these two tendencies?—a balance between innovation and tradition? All of these are open questions, among many other open questions, of central importance and still current today. The ancient Greeks were the first to ask some of them, and these questions re-emerge in modern-day universities and social science. We will deal with them in the following chapters. Nevertheless, one general observation regarding ancient Greek higher education might be the tendency to primarily teach values and cultivate virtues, rather than confer vocational competence that permitted its pupils to face the material challenges posed by life. Like the paths to culture and to erudition, these paths, too, can integrate with each other, just as they can obstruct each other. The purpose of a good school should be to find a balance between these tendencies.

The reader may feel, at this point, somewhat confused. And, indeed, our contemporary higher learning system and institutions grew out of many traditions, at times contradictory. One of the purposes of this book is, if not to bring clarity, to awaken consciousness to this confusion.

CONCLUDING REMARKS

- Higher education was, from ancient time, practiced by clerics and strictly interconnected with religion.
- Higher education in the ancient world was primarily, although not exclusively, vocational. Non-vocational teaching wisdom gained impetus

with the dawn of Greek philosophy. These two different tendencies are, as we shall see, at the root of confusion in the modern university.

- The method of study in ancient time was essentially mnemonic, and this gave rise to learning that still today is more acquisition of data than formative.
- The pre-Socratics asked fundamental questions about knowledge that are still open today. The attempts to answer these questions are occasionally related to conflicting views, such as traditionalism and progressivism, or individualism and collectivism. We find these tendencies in higher education, both ancient and modern.
- As early as in ancient time, knowledge was related to social rank, a relation that today can degenerate to a misuse of knowledge.

NOTES

1. *The Life of Pythagoras*, http://www.tertullian.org/fathers/porphyry_life_of_pythagoras_02_text.htm, accessed May 11, 2014.
2. The assertion that computers cannot fully replace a live teacher is a part of the thesis developed by Joseph Weizenbaum in his classic *Computer Power and Human Reason* (1976), regarding the impossibility of the computer to surpass human reasoning.
3. Gilbert Ryle, in his article "The Academy and Dialectic," in Ryle 1971, pp. 89–125, discusses the various types of dialectics and their origins.

2 The Classical Roots
Farewell to the Socratic Method

Socrates formed the spirit

Already in ancient times, learning based on dialogue and critique—perhaps the most appropriate for scientific studies—was, for socio-political reasons, replaced, albeit gradually, by a unidirectional and authoritarian teaching. Moreover, practices that are alien to science and study, such as applying pressure and causing stress, were introduced.

The most famous teacher of the dialectic tradition was Socrates, who lived at the end of the 5th century BC in Athens. Although surrounded by disciples, he did not leave any writings, nor did he create a school of his own. Much has been said and written on Socrates, but very little is known about him for certain, and the history of philosophy warns against relying excessively on the available sources. One important such source are the writings of Socrates's pupil, Plato (c. 429–347 BC), and the latter's interpretation of Socrates's dialectic as, inter alia, philosopher Gilbert Ryle (1900–1976) points out, is at times contradictory.[1] One other source is that of Socrates's other pupil, Xenophon (c. 430–354 BC), who portrays Socrates differently than Plato. Both sources are unreliable. Popper, too, in his admiration for Socrates, may be painting an excessively idealistic portrait of the latter (*OS* 1, 128–133). Thus, we only have a very general knowledge of Socratic thought and his method, known as "maieutics" (a term that derives from *maia*—mother, midwife), which Plato, as we will see and as Popper propounds, probably put an end to.

In broad terms, the Socratic method consists of a dialogue in which a particular question is dealt with by using critique as a stimulus to increasingly better answers; more specifically, one interlocutor leads another to contradict himself, realize his error, produce new conjectures—and grow. There are no limitations, and because no final answer might ever be reached, the game (like scientific research itself) can becomes endless (Popper's intellectual biography is titled accordingly *Unended Quest.*).[2] Socrates most probably presented himself humbly, as an ignorant person, thus placing his interlocutor in the position of a sham teacher, encouraging him to define his own position. The Socratic method aims at helping people to learn to think, examine themselves, form their character and improve themselves by correcting errors. It has no pretense to imparting knowledge, although it can greatly facilitate access to knowledge. According to Plato, and in particular to the dialogue *Theages*, attributed to Plato, Socrates was well aware of the difference between his interactive method and the unidirectional transfer of knowledge.

The Socratic method in the form described previously was only sporadically applied over the course of history and, with few exceptions, was little employed by universities. Nevertheless, those who have tried it could experience the extent to which it encourages creativity, clarity, sincerity, and intellectual honesty, and how enthralling it can be. Because it consists of a dialogue, it can be applied to only a small number of students at a time— ideally just one. Therefore, it can become expensive, perhaps even elitist and unsuitable for the masses, even though it could be applied with positive results in mass teaching institutions, which is what many modern universities are today. Above all, in its constant pursuit of truth, it does not claim truth for itself, nor does it become dogmatic. Naturally, it must be applied with caution and a respectful etiquette must be followed in order to keep the debate from getting out of hand. The most insidious effect, for some members of an authoritarian society, is that it helps the interlocutor to learn to think with his own head and become independent. No wonder Socrates was condemned to death.

Whereas Socrates, without school, is said to have spent his days conversing everywhere, with anyone and on any topic, the merit for having founded classical antiquity's earliest secondary schools goes to the next generation of Athenian teachers, in particular to Isocrates and Plato (5th/4th centuries BC). Theirs were not the only contemporary Greek schools of higher learning, but they are the best documented. Admittedly, they are known above all through the fame and works of their founders and not through reliable descriptions of their functioning; we do not know to what extent their opinions concerning education were implemented in their own, or in other, institutions. Their philosophy, nonetheless, influenced profoundly later higher learning and indicates that, despite their respect for their teacher Socrates, they had deviated from his maieutic.

The earliest higher schools, founded by Isocrates and Plato, inaugurated the transfer of knowledge

Greek higher education consisted in learning the *liberal arts*. To be sure, the term appears first as late as in the 1st century AD, in Seneca's *Moral letters to Lucilius (Letter 88)*: "*De liberalibus studiis.*" They were those subjects or skills considered essential for a free man (who did not need to practice servile manual arts) to know in order to take an active part in civic life. The liberal arts later came to be seven (see next chapter): grammar, dialectic, rhetoric, geometry, arithmetic, astronomy, and music (the latter consisting usually in learning the literary tradition handed down largely in song and dance and in similar rituals), which were to play such an important role in university studies; and from which many branches of modern science developed.

The liberal arts, or at least some of them, were taught in Isocrates's and Plato's schools. Isocrates (436–338 BC), an admirer of Socrates, founded a school in Athens with a teaching that went beyond basic knowledge. In ancient times, and in medieval universities, instruction was not separated into primary, secondary, and higher education as it is today; for many centuries, the dividing line between the various levels of studies remained blurred. Isocrates's school provided a cycle of studies of three or four years and was open to all, unlike other institutions, such as the Pythagorean sect or Plato's Academy (which we will soon discuss), and charged a fee. At Isocrates's school, students learned and practiced gymnastics, music, grammar, literature, history, mathematics, dialectic, and, above all, the art of oratory, but not for the same reasons as those of the Sophist tradition. Isocrates was critical of education based only on rhetoric, such as eristic, whose sole goal was to best one's adversary. He introduced values, notions, and practical aspects; in other words, knowledge. The historian of classical education Henri-Irénée Marrou (1904–1977), while reminding us that "Isocrates is not in the same street as Plato," does admit that "It is to Isocrates more than to any other person that the honour and responsibility belong of having inspired in our Western traditional education a predominantly literary tone" (Marrou 1982, 79–80). Isocrates's school produced successful politicians and was a rival of the other, more famous school that applied elitist philosophy: Plato's Academy.

The biographer of Greek philosophers, Diogenes Laertius, who probably lived in the 3rd century AD, narrates (*DL* III/7) that Plato "lived in the Academy, which is a gymnasium outside the walls, in a grove named after a certain hero, Hecademus": hence the term *academy*. The "gymnasium," which derives from the Greek *gymnasion*, "gymnastic school," was a facility where young men trained for athletic competitions. Later, it became a place for socializing and intellectual activities, and later still, in the Hellenistic era, that is, the period that began with Alexander the Great's conquests, it evolved into a true school. Today, in many countries, secondary schools are often called "gymnasium." Here, Plato established both his school and his personal abode; a number of his pupils most likely lived there as well. The site of Academy was sacred to Athena, the goddess of wisdom, and other immortals. Plato is also said to have erected a temple dedicated to the muses on the site (*DL* IV/1). If Plato was not a cleric, he behaved like one. Without entering here into the debate concerning Plato's view of religion, his Academy, like almost every institution of higher learning in antiquity, to repeat, was associated with religion, a characteristic that, as we will see, later accompanied and influenced, in form or in substance, higher education up until the French Revolution. Moreover, even if the university as an institution was officially secularized in the 19th century, still today it continues to be burdened with dogmatic elements reminiscent of religion.

In general, it is not quite clear how the Academy functioned under the guidance of its founder.[3] It appears not to have charged fees, although Plato did not disdain presents. Teaching took the form of a debate, although for different purposes than the Socratic debate, and a preferred topic was mathematics. Tradition has it that an inscription above the Academy's door warned, "Let no one ignorant of geometry come under my roof!"[4] Among Plato's pupils one could find women as well (*DL* IV/1).

Plato's Academy, too, had a considerable influence on modern higher education. Just consider how widespread the term *academy* has become to indicate various types of educational, cultural, or scientific institutions, be they of high level or aspiring to such. The Academy continued to function (although, as it seems, with interruptions) for centuries. The circumstances of its dissolution are not certain. In 529 AD Byzantine Emperor Justinian (r. 527–565) banned pagan teaching throughout his empire, and histories repeatedly imply that the Platonic Academy was among the schools that had to cease their activity. No doubt the gradual rise of Christianity played a determinant role. There is, nonetheless, no evidence for this, and the Academy may have well survived till the Slavonic invasion around 580.[5]

In addition to its religious orientation, Plato's Academy did not teach professions, nor did it confer any title. But the same can be said of Isocrates's school and other higher schools of classical antiquity. The goal of Platonic education was sociopolitical: to form better men, able to guide the state.

How are political leaders formed? How is an elite formed? The two categories are not necessarily equivalent; the first denotes a guide; the second, a group of privileged individuals. Nevertheless, Plato, in his attempt to deal with the problem of social stability, does not seem to distinguish between the two and, hence, another ambiguity that is still very current today.

Plato was a pessimist; he feared that society could degenerate, and in order to avoid this he felt that continuous vigilance was needed. But training custodians also meant running the risk of creating independent individuals and encouraging new ideas and change. In order to ensure that the future leaders towed the line, and that they made their underlings tow it as well, he suggested a rigorous selection process, which calls to mind the selection process of various modern universities, at times called "centers of excellence," although they often encourage conformism and even mediocrity, rather than excellence. We will soon return to this topic.

In *The Republic*, one of his best known works, Plato presents a utopian design for maintaining the stability of society with an educational plan that is reserved to the upper classes. To Plato, the ideal state is structured like a pyramid composed of three classes or castes, each with its own rights and duties. At the top are the governing philosophers; at a lower level are the warriors whose task is to defend the state; at the base is a multitude of workers, such as farmers, artisans, and other subjects who uphold the economic structure. And then there are the slaves, who do not count as a class. Each class has its own role that it carries out according to specific training, and this keeps the social pyramid stable.

Plato introduces stress into study

The efforts required to maintain the society's stability should have fallen almost entirely on the shoulders of the philosophers and the warriors. The initial phases of the two classes' education were basically similar and consisted of gymnastics and music, this latter to train the soul. Already at this stage, Plato suggests imposing "toils, pains and conflicts" (*The Republic* III, 413 d).[6]

Are toil, pain, and competition necessary in studying? One of the aims of this book is to suggest that they are not only unnecessary but also counterproductive. Toil is a continuous and disagreeable activity that risks diminishing intellectual ability, and pain could alienate interest in the subject being studied. The "conflicts" were introduced most likely because of the Academy's context as a "gymnasium" with athletic competitions. Moreover, a competition implies a specific goal and, thus, limits the possible range of growth of knowledge to one single direction. Yet if a student studies out of pure interest and not out of direct or indirect imposition, the exertion becomes negligible. Institutions of higher learning can indeed exist in which people study for sheer pleasure, and, moreover, one can apply methods of study that minimize the pain, such as the Socratic method. And yet, still today, it is obvious that students cannot learn without toil, pain, and conflicts. Why? Because study and science are often coupled with didactic procedures and goals that are not in harmony with the subject studied.

The combination of toil/pain/conflicts in study has, thus, deep historical roots and might have been introduced, at least on a theoretical level, by Plato, in his pessimistic assumption that human beings and society have a propensity to degenerate and, thus, need to be subjected to strict control. The result is an authoritarian system in which obstacles are chosen as a function of a given orientation in order to select the most conformist candidates—what one calls today *meritocracy,* that is, a social class that best responds to the ruler's request.

Meritocracy is a term coined in 1958 by the British sociologist Michael Young (1915–2002), although the concept was well known in ancient times. In a small book titled *The Rise of the Meritocracy* (1970), Young portrays and criticizes Britain under the future rule of a class favoring merit. *Merit* can nevertheless mean or depend on different features, such as intelligence, competence, credentials, commitment, popularity, and, above all, success— all assessed in relation to a framework. Merit is often confused with excellence, which—as the Latin root of the term suggests—exceeds accepted norms. Meritocrats can be excellent individuals, yet their performances are normally judged in relation to specific frameworks and goals; as such, they conform; the excellent often do not. Hence, the propensity of mediocrity, so characteristic to meritocracy. The Platonic goal of "toil, pain and conflicts" is to select future leaders by breaking the backs of the "weaker" candidates, incapable of being part of the herd even if they are often potential sources of new ideas that may be excellent. Hard and painful training might perhaps be useful in forming warriors, but are they truly needed to form intellectuals, humanists, or scientists?

Plato frankly recognizes the suffering and damage that such study can cause—a salient detail generally omitted nowadays: "For the mind more often faints from the severity of study than from the severity of gymnastics" (*The Republic* VII, 535 *b*). He warns, "Knowledge which is acquired under compulsion obtains no hold on the mind. . . . Do not use compulsion, but let early education be a sort of amusement; you will then be better able to find out the natural bent. . . . For sleep and exercise are unpropitious to learning" (*The Republic* VII, 536 *e* and 537 *b*). Unfortunately, this contradicts his own request to impose "toil, pain and conflicts" and neglects the inevitable competitions that the classist society sparked by playing on personal ambitions. Already in Plato's writing we find the contradiction, as evident as it is today ignored.

An interesting detail: Plato believed that women, too, could belong to the caste of warriors, an unusual opinion for his day, as he himself admits in the Fifth Book of *The Republic*, "Many more doubts arise about this than about our previous conclusions" (*The Republic* V, 450 *c*). Nevertheless, he affirms, "Then, if women are to have the same duties as men, they must have the same nurture and education" (*The Republic* V, 451 *e*–452 *a*). More precisely, Plato also proposes a "community of women and children" (*The Republic* V, 450 *c*). In other words, women and children are members of the caste and its belongings—an unusual combination, the fruit of ancient customs and (possible) progressive philosophical aspirations. Plato's affirmations regarding the education of women, despite their brevity, have been the object of many interpretations and much discussion; but perhaps he was merely influenced by the martial talents of Spartan women.

In more detail, this is how leaders are to be formed according to Plato's design. At twenty years of age, candidates undergo the first selection process; the worthy students continue their studies of the (technical) sciences: "numbers and calculations" (i.e., arithmetic), geometry, astronomy and harmony. Those who study these subjects, which are also useful on the battlefield, are ready to become warriors. "And he who at every age, as boy and youth and in mature life, has come out of the trial victorious and pure, shall be appointed a ruler and guardian of the State; he shall be honored in life and death. . . . But him who fails, we must reject" (*The Republic* III, 413 *e*–414 *a*).

Nevertheless, to Plato, the aforementioned "sciences" are only propaedeutic for the study of true knowledge. Real instruction, reserved to the future leaders, begins at thirty years of age; later, another selection launches the worthy in the study of dialectic, which lasts roughly five years and suppos- edly develops the faculties needed to achieve awareness of truth and success in politics. At fifty years of age, "let those who *still survive* [*sic*] . . . making philosophy their chief pursuit, but, when their turn comes, toiling also at politics." (*The Republic* VII, 540, *a* and *b;* italics mine).

Dialectic is then the pinnacle knowledge, and access to this field of phi- losophy is strictly reserved to those people who have demonstrated their loyalty. The technical knowledge needed for the requirements of daily life is less important. This includes the ability and courage of the warriors to defend the state and, less importantly, the activities of the workers who, thanks to their temperance, guarantee the livelihood of society. Closing the ranks is manual labor, which was in Plato's day considered a contemptible activity to be relegated to the slaves. Here is what Plato has to say to this regard:

> The finest and most suitable kinds of learning are those which will bring him the most reputation for philosophy; and he will get most reputation if he appears well versed in all the arts, or if not in all, in as many of them, and those the most considerable, as he can, by learning so much of them as befits a free man to learn, that is, what belongs to the under- standing rather than the handiwork of each.
>
> (Plato, *Lovers*, 135 *b*)[7]

Thus, Plato, who favored an elitist society, established a relationship between knowledge and social classes, using a scale that is not necessarily valid, but is nonetheless still present today. At the apex is the teaching that develops the virtues; at the base is the humble manual labor carried out by the slaves.

Within the Platonic framework, the cultured and free man does not work; therefore, study is not vocational. Yet establishing a priority among the various subjects and placing them in relation to the social classes distances Plato from the Socratic principle of continuous and unconditional critique and improvement. Plato began to distance himself from his teacher in his own didactics. Moreover, Plato's method was no longer merely formative; it was also the beginning of an acquisition of knowledge.

From formative learning to acquisition of knowledge

At the end of the Sixth Book of *The Republic*, Plato presents his "line theory"; an imaginary line is divided into four segments that represent the four types of objects with which the soul comes into contact. Two seg- ments represent the intelligible world: first, philosophy, "the true science"; then intelligible truth based on observation, such as geometry. And the other two represent the visible world of visible objects and images.

And when I speak of the other division of the intelligible, you will understand me to speak of that other sort of knowledge which reason herself attains by the power of dialectic, using the hypotheses not as first principles, but only as hypotheses—that is to say, as steps and points of departure into a world which is above hypotheses, in order that she may soar beyond them to the first principle of the whole.

(*The Republic* VI, 510 *b*)[8]

As opposed to Socrates, Plato claims for the ability to soar the "first principle of the whole"—of truth. In today's higher education as well—and not only there—it is believed, at least subconsciously, that one is teaching, or acquiring, truth. And this, as we will see, is a grave error.

How does one achieve the truth? Plato is not very clear on this point. Dialectic is useful only for the first part of the journey, "as steps and points of departure into a world which is above hypotheses . . . to the first principle of the whole" (*The Republic* VI, 510 *b*). It helps to cleanse the soul of "falseness." In his dialogue *Meno*, Plato holds that knowledge can be re-evoked through recollections of past lives. Thus, even a slave, if stimulated through questions, can manage to solve a geometry problem without ever having studied the subject.

And yet, truth is achieved, at least partially, through illumination, a process that is metaphorically described at the beginning of the Seventh Book of *The Republic* by means of the famous allegory of the cave. Plato describes men imprisoned in a cave; one of them—who will become a philosopher—manages to free himself, go outside, be illuminated, and "compel the best minds to attain that knowledge which we have already shown to be the greatest of all" (*The Republic* VII, 519, *c–d*): he imposes the truth on those who have not yet been illuminated. In the end, Platonic education, despite the importance he gives to dialectic, gradually turns into transfer of knowledge; a chosen person is initiated to propagate to (or better, to impose on) ignorant people true knowledge, originating from who knows which higher source.

A fundamental issue that Plato neglects is how can one be sure that illumination is genuine? The answer, if it exists, transcends philosophy. According to the Old Testament, for example, the law is conceded directly from God to Moses, who then transmits it to the people of Israel. Nevertheless, the context is that of faith, and the Bible warns against false prophets. In rational philosophical thought and in science, how can one be sure that the illumination is genuine and leads to the truth?

In any case, what we have is a change in teaching methods, on both a methodological and a social level, and in the relationship between teacher and pupil. In the Socratic method, knowledge grows through continuous dialectical interaction, in which the teacher also accepts a passive role and the pupil may assume an active role. With Plato, knowledge is a priori, and the pupil becomes passive and subjugated, whereas the teacher is active and authoritarian; knowledge grows deductively from fundamental ideas and is transmitted in unidirectionally.

The goal of the Platonic social program is to maintain social stability. Fortunately, it was never put into practice, even if history has, regrettably, witnessed regimes that have done worse. In the past century, in the shadow of two dictatorships, Popper, like Plato, asked himself how social stability could be maintained, well aware of the need to balance tradition and progress and to minimize harm therefrom. Popper harshly criticizes Plato's solution as a source of totalitarian thinking that immobilizes society. According to Popper, Plato's educational system transforms knowledge into a means for furthering one's career, an obstacle course in which the student strives to acquire useful knowledge, not in order to grow intellectually, but merely for social advancement:

> This tendency transforms our educational system into a race-course, and turns a course of studies into a hurdle-race. Instead of encouraging the student to devote himself to his studies for the sake of studying, instead of encouraging in him a real love for his subject and for inquiry, he is encouraged to study for the sake of his personal career; he is led to acquire only such knowledge as is serviceable in getting him over the hurdles which he must clear for the sake of his advancement.
>
> (*OS* 1, 135)

Who, among today's university graduates, hasn't experienced this distress first-hand?

Popper identifies Plato as the first theoretician of the "closed society," that is, the dogmatic and tribal society; as such, Plato is also the first great enemy of the "open society"—the kind of society that encourages criticism. This latter type of society is, according to Popper, the most appropriate kind for the growth of science and democracy. The central argument of the present book is that even though the modern university monopolizes science, it remains, for historical reasons, still a closed society in many ways, subject to tensions that are in part vented upon those who are its components, above all on students, causing them useless and counterproductive stress. As to the Academy under Plato, as classicist Ian Mueller (1992, 171) remarks, the educational timetable of *The Republic* seems totally impracticable for a privately organized institution in a free city. Popper adds the following:

> Plato did not prove too successful as a selector of political leaders. I have in mind not so much the disappointing outcome of his experiment with Dionysius the Younger, tyrant of Syracuse, but rather the participation of Plato's Academy in Dio's successful expedition against

Dionysius. Plato's famous friend Dio was supported in this adventure by a number of members of Plato's Academy. One of them was Callippus, who became Dio's most trusted comrade. After Dio had made himself tyrant of Syracuse he ordered Heraclides, his ally (and perhaps his rival), to be murdered. Shortly afterwards he was himself murdered by Callippus who usurped the tyranny, which he lost after thirteen months. (He was, in turn, murdered by the Pythagorean philosopher Leptines.) But this event was not the only one of its kind in Plato's career as a teacher. Clearchus, one of Plato's (and of Isocrates') disciples, made himself tyrant of Heraclea after having posed as a democratic leader. He was murdered by his relation, Chion, another member of Plato's Academy. (We cannot know how Chion, whom some represent as an idealist, would have developed, since he was soon killed.)

(*OS* 1, 136)

Not all what Popper describes must have been an outcome of Plato's legacy, yet *The Republic*, along with Plato's other works, offers a vivid idea of the aspirations and values of his time and helps to understand some of the successive intellectual developments, even if not all of Plato's disciples followed his teachings. Plato's most famous pupil was Aristotle, who spent twenty years at the Academy. Before we deal with him, let us summarize a few points regarding higher education that the ancient Greeks passed down to us:

CONCLUDING REMARKS

- Starting with Socrates—and perhaps even before him—the ancient Greeks already distinguished between formative learning and plain acquisition of knowledge. From Plato on, teaching gradually evolved from the former towards the latter.
- The goal of classical education was not vocational (although it offered considerable advantages, in particular for those endeavoring to become politicians); it used dialectic, albeit in different ways:
  - Socrates used it to correct errors, although he never claimed to know the truth in advance.
  - Plato claimed to know the truth all along; he used dialectic to contrast false ideas. And yet, knowledge (*alias* truth):
    - Is "a priori," or rather, details are deducted from general principals.
    - Is transmitted in a unidirectional manner.
- Problem: education and dialectic cultivate personal abilities, which could develop in an uncontrollable and harmful manner, thus compromising social stability.

- Plato's solution: to select the candidates to be educated so as to ensure their loyalty. However, this creates other problems:

  - Rather than concentrating on developing their virtues, the candidates concentrate on finding ways to overcome the obstacles created by the system.
  - Study becomes distressing and students may lose interest in the subject.
  - It creates an elite, an exclusive and privileged group, and excludes from study non-conformists who could otherwise contribute freely to society and progress.
  - It hinders progress and fosters stagnation.
  - An elitist method of teaching, such as Plato's, induces people to lie; indeed, according to Plato, the philosophers who lead the state have the right, and indeed the responsibility, to lie.

BIBLIOGRAPHICAL NOTES

A fundamental text for the study of Greek philosophy is *Lives of Eminent Philosophers* by Diogenes Laertius, in which the author presents in anecdotal form the lives of 83 ancient thinkers and their appartenance to philosophical schools. Almost nothing is known about Diogenes Laertius, who is believed to have lived between the end of the 2nd century and the first half of the 3rd century AD.

Besides Diogenes Laertius's *Lives,* we have, of course, the corpus of works by Plato; *The Republic*, which was written by Plato as a dialogue around 390 BC, being particularly pertinent, as it presents the author's view of the utopian state.

There are, of course, many other sources and studies. One important study is Werner Jaeger's *Paideia*: a monumental work, in three volumes, published for the first time in German between 1934 and 1947. *Paideia* means "education" or "culture," and the work studies Greek literature and philosophy in detail, attempting to reconstruct the various aspects of education in Greek civilization. Another general work, which is occasionally useful—although historiographically dated, is *A History of Education in Antiquity* by Marrou, published initially in French in 1948. The title notwithstanding, the book concentrates upon the classical world, neglecting other civilizations or ancient cultures and, anyhow, does not deal specifically with higher education.

A recent, clear, and rigorous study of Plato's Academy is John Dillon's, *The Heirs of Plato* (2003). In the first chapter, he discusses the sources that deal with the location and the functioning of the Academy in Plato's time; he criticizes Cherniss's *The Riddle of the Early Academy* (1945) as speculative and imprecise. The first two introductory chapters of John Patrick Lynch *Aristotle's School: A Study of a Greek Educational Institution* offer an outline of higher education in Athens in Plato's day.

NOTES

1. Ryle 1971, 100. To Hadot 1998, every philosopher can create his own personal Socrates, even if Hadot keeps his distance from the rational interpretation of Socratic thought when considering philosophers such as Kierkegaard and Nietzsche.
2. Popper 1976.
3. For a brief discussion see Mueller 1992, 170–75, and for more details see Dillon 2003, ch. 1.
4. See Suzanne, http://plato-dialogues.org/faq/faq009.htm, accessed May 12, 2014.
5. For a detailed discussion see Lynch 1972, ch. VI. Lynch remarks, inter alia, that in Justinian's *codex* there is no related call, and that some of the teachers of the school went on teaching after 529.
6. Quotations are from www.gutenberg.org/files/1497/1497-h/1497-h.htm, accessed August 9, 2011.
7. http://www.perseus.tufts.edu/hopper/text?doc=Perseus%3Atext%3A1999.01.0176%3Atext%3DLovers%3Asection%3D135b, accessed August 15, 2011 (Source: Plato 1955).
8. Quotations from http://www.gutenberg.org/files/1497/1497-h/1497-h.htm#2H_4_0009, accessed on August 16, 2011.

3 The Classical Roots
Aristotle and Beyond

Aristotle: "the first to teach and write like a Professor"

Aristotle's teaching is most relevant to the history of the growth and transfer of knowledge. Bertrand Russell (1872–1970) remarks that Aristotle "is the first to write like a professor." (Russell 1975, 174). With him, knowledge expands and acquires methods, some of which still persist today: the role of dialectics contracts and is largely replaced by observation—all practices relatively common in modern higher education. Who was Aristotle and what was his contribution, for better or worse, to knowledge and higher learning? To be sure, Aristotle's learning is complex, at times contradictory, and has been subject to many interpretations, past and present.

Aristotle was born around 384 BC in Stagira, Macedonia, and like other philosophers and scientists we will encounter, he was, inter alia, associated with a court. His father was a friend and the physician of Amyntas II, king of Macedonia (d. 393 BC). Aristotle studied and taught for twenty years at Plato's Academy. He then migrated for a number of years, and in 343 joined the court of King Philip II of Macedonia (r. 359–336 BC) and tutored his son, Alexander "the Great" (356–323 BC). Tutoring was the most common way of teaching in higher standing families till modern times.

Diogenes Laertius relates that Aristotle, returning to Athens, aspired to run Plato's Academy but found that the position was occupied by another disciple of Plato and, as a result, he founded his own school:

[. . .] while he was absent on an embassy to Philip, on behalf of the Athenians, Xenocrates became the president of the school in the Academy; and that when he returned and saw the school under the presidency of someone else, he selected a promenade in the Lyceum, in which he used to walk up and down with his disciples, discussing subjects of philosophy till the time for anointing themselves came; on which account he was called a Peripatetic. But others say that he got this name because once when Alexander was walking about after recovering from a sickness, he accompanied him and kept conversing with him.[1]

(*DL* V, 2–3)

Aristotle established his school in a public gymnasium dedicated to learning and—as in the case of Plato's Academy—of worship, consecrated to Apollo Lyceus (Apollo the "wolf-god"—hence the name "Lyceum") and to the muses. Aristotle, too, acted as a priest, although grudgingly. He may have endeavored to emulate Plato and count as an arch-philosopher (although this expression is an oxymoron because philosophy knows no authority). His gymnasium was also known as "Peripatos"—colonnade. "Peripatetic" indicates likewise, as suggested by Diogenes Laertius, walking around while discussing philosophy.

Aristotle taught at the Lyceum for twelve years. In 323 BC, soon after Alexander's death, Athens reacted to the Macedonian rule. Being associated with the Macedonians, Aristotle, a year later, had to flee the city. The guidance of the Lyceum passed to his pupil Theophrastus (c. 371–287 BC), and Diogenes Laertius (*DL* V) lists a sequence of philosophers who subsequently ran the Lyceum. The school eventually declined, although it seems to have survived the fall of Athens to the Romans in 86 BC. As in the case of Plato's Academy, we do not know for certain when exactly it ceased to exist.

Relatively little is known about the Lyceum under Aristotle's guidance. Like Plato, Aristotle does not seem to have taken fees, and, in addition to discussing philosophical questions with his pupils, he would teach publicly on a more popular level. Aristotle's school, and perhaps Plato's, was in any case, less sectarian than the Pythagoreans who kept their knowledge secret—a progress towards modernity. The immense body of Aristotelian writings that has survived, encompassing morality, aesthetics, logic, metaphysics, politics, and science, is to a large degree formed by Aristotelian lessons transcribed by his pupils. When speaking of Aristotle's "science," let us, however, not confuse ancient concepts with modern ones. For example, Aristotle's *physics* or *meteorology* are very different fields than those that bear these names today. "Physics" (from φύσις: *physis*) can be translated as "nature" and was concerned with the internal activity that makes anything what it is. "Meteorology," (from μετέωρος: *metéōros* lofty, high) meaning a discourse from high in the air, was concerned with earth sciences and had a much broader sense than today. Both were qualitative fields, not quantitative as they are today.

Generally speaking, Aristotle pictured the world as a nest of spheres all centered about the earth and carrying the sun, the moon, and the five planets known at that time. He conjectured that different laws governed the celestial and terrestrial ("sublunary") regions. He adopted Empedocles's view that matter on earth was composed of four basic elements. Each of the elements had a "natural tendency"—water and earth toward the center of the universe, air and fire away from it—and these tendencies determined the physical behavior of a substance. Aristotle denied the existence of vacuum and of atoms and held that matter is continuous. From a modern and fairly unhistorical point of view, however, one could say that Aristotle's main contribution to science was the study and the classification of living beings. There are, to repeat, many interpretations of Aristotle and his school, and, thus, the brief outline in this chapter is inevitably oversimplified, incomplete, and open to debate.

With Aristotle, observation becomes the main source of knowledge; as a result, the acquisition of knowledge becomes inductive and knowledge itself cumulative and, occasionally, dogmatic

How should one learn according to Aristotle? To be sure, Aristotle offered more than one theory of learning, or at least his theory of learning may be interpreted in different ways. Simon Blackburn's excellent *Oxford Dictionary of Philosophy* presents the traditional view that places Aristotle essentially in contrast to Plato:

> The traditional contrast is between Plato's otherworldly, formal, and *a priori* conception of true knowledge (*noēsis*), as opposed to Aristotle's intense concern for the observed detail of natural phenomena, including those of thought, language, and psychology.
>
> (Blackburn 1994, 25)

Not everybody agrees, and a deeper study of Aristotle reveals different nuances. Popper (*OS*, chap. 11) views Aristotle as closer to Plato. Indeed both took learning to be an intuitive, ecstatic experience (though Plato set ideas outside space and time, and Aristotle disagreed; this disagreement, however, is metaphysical and so irrelevant in our context). Moreover, both took dialectic to be not learning but a mere preliminary to learning.

Let me, however, for simplicity, return to the traditional and perhaps crude view of Aristotle, which is also relevant to later developments in higher education. Aristotle was well aware that the teachings of Plato were not feasible, at least not in the near future. He proposed a temporary compromise, which became the basis of Western learning for the next two thousand years. As in Plato's case, however, Aristotle could hardly have wholly applied his theory of learning.

Aristotle distinguished (e.g., in his methodological work *Sophistical Refutations*, 165 *a–b*) between didactic and dialectic arguments. *Didactic* arguments are "those that reason from the principles appropriate to each subject and not from the opinions held by the answerer (for the learner should take things on trust)";—that is, the transfer of data. *Dialectic* arguments are "those that reason from premises generally accepted, to the contradictory of a given thesis"—aimed, inter alia at the personal development of the individual.[2] In Aristotelian learning, too, one finds a combination of dialectic and the unidirectional transmission of knowledge, but in the opposite order compared to Plato's teaching. Whereas Platonic learning begins with dialectic as a preparation for receiving truth, Aristotle preferred various indisputable principles obtained through observation or common sense:

> If our reasoning aims at gaining credence and so is merely dialectical, it is obvious that we have only to see that our inference is based on premises as credible as possible: so that if a middle term between A and B is credible though not real, one can reason through it and complete a dialectical syllogism. If, however, one is aiming at truth, one must be guided by the real connections of subjects and attributes.[3]
>
> (*Posterior Analytics* I, 19, 82*b*)

Aristotle, then, favors seeking and accumulating observed data. Later, dialectics may be used in reply to the criticism aimed at these principles.

Both Aristotle and Plato believed that true knowledge can be reached, although by entirely different methods. For Plato true knowledge lies in ideas, whereas Aristotle believed, at least according to the traditional view, that knowledge is acquired primarily through observation or common sense, indeed, by induction.[4] The growth of learning separates into two traditions. In the first, which is deductive "a priori," one begins with basic principles (which could be, inter alia, dogmas) to then arrive at the details. In the second, which is inductive "a posteriori," one begins with observation to then formulate general principles. This latter tradition is more laborious than the former and calls for perfectionism in the collection of details, in turn at the risk of pedantry, thence leading to authoritarianism and, again, dogmatism. It will be considered dominant in modern science and in the related philosophy of science, even though, as we will later see, these arose in the 17th century as a reaction to the predominance of Aristotelian thought itself. Aristotle teaching became, in any case, more informative and erudite than formative.

An aspect of the cultivation of inductive knowledge is the belief that it accumulates as though it were a liquid in a container. This results today in a series of practices such as trying to measure knowledge, for example, on the basis of the number of credits offered by a university course, the number of pages studied from a textbook, or the length of a thesis, an article, or a book. According to this criterion, Galileo's *Sidereus Nuncius*, the small book numbering about sixty pages, including illustrations, which in 1610 inaugurated modern astronomy by presenting sensational telescopic discoveries, would not have much of a chance; and even less so the 30-odd-page article in which Einstein laid the basis for the Theory of Relativity in 1905. Nevertheless, knowledge, as Popper argues, is subject to different growth mechanisms: rather than accumulating, it evolves; in his *Objective Knowledge* (1972), Popper criticizes the concept of the mind as a passive receptacle of knowledge.

Aristotle (and his followers) rightly favored an approach that stresses the unity of knowledge. However, in trying to achieve this, he encouraged the opposite: the systematization of knowledge resulting in its modern fragmentation.

Aristotle inaugurates the classification of knowledge

Like Plato, Aristotle prioritized knowledge, a ranking that also became decisive for subjects that were to be studied at the university:

1. At the top of the ranking was *theoretical knowledge*; that is, the sciences that seek knowledge for its own sake, including metaphysics, physics, and mathematics. As opposed to the Pythagoreans and Plato, however, Aristotle underestimated the importance of mathematics at the expense of a qualitative study of the essence, an attitude that, in the Middle Ages, helped bog down the development of Western sciences.
2. Next came *practical knowledge* (not to be confused with technical knowledge), which seeks moral perfection and includes ethics and politics. Aristotle did not consider history a science because it is not universal but only of the particular.
3. Lastly, *poetic knowledge* ("poiesis" is the creative moment of the spirit), which deals with production: the fine arts and *techniques,* such as architecture or weaving.

To Aristotle (*Politics* 8.3, 1337 *b*), much like Plato, the professions that "deform the body"—that is, manual or paid labor—are vulgar "for they absorb and degrade the mind."

> [. . .] but leisure is better than occupation and is its end; and therefore the question must be asked, what ought we to do when at leisure? [. . .] It is clear then that there are branches of learning and education which we must study merely with a view to leisure spent in intellectual activity, and these are to be valued for their own sake; whereas those kinds of knowledge which are useful in business are to be deemed necessary, and exist for the sake of other things.[5]

The above scale of priorities, looking down on manual and technical labor, continues to be current, at least by implication: theoretical knowledge is still considered more prestigious than applied or practical knowledge, and a white collar worker is more appreciated than a blue collar worker, even when the blue collar worker actually makes more money! All these distinctions derive more from tradition than from actual contributions to knowledge in whichever field.

What should be the aim of education? Aristotle was less elitist than Plato, and rather than preparing leaders for subdued citizens, he aspired to prepare autonomous citizens able to choose their leaders on their own. By "citizens," Aristotle—as opposed to Plato—intended only free adult Greek males. He considered women inferior beings and had no place for them in his educational system.

But if citizens are autonomous, not subject to control, how can they be kept from doing damage? This question was not posed by Aristotle but considered some time later by pessimist educators who assumed, as Plato did, that one is by nature more inclined to do harm than good. Aristotle's classification offered, however, a way out: to transmit knowledge in airtight compartments; the more individuals learn and develop—then and now— the more they concentrate on specific aspects, possibly neglecting context and criticism. This encourages concentrating on one's own field and avoiding interfering in the fields of others to the detriment of interdisciplinarity, which is one of the driving forces of science. Today, interdisciplinary subjects, such as biophysics, the history and philosophy of science, and many others, have contributed greatly to the evolution of knowledge and have become independent disciplines. Biophysics, in particular, would never have been admitted into the Aristotelian disciplinary structure because— without going into the details of Aristotelian classifications of living beings— it would not allow the study of the living to cross over into the study of the non-living. The tendency to limit interdisciplinarity still persists today. To limit interdisciplinarity means to limit science.

Aristotle dealt with the problem of which subjects to learn

Whereas Plato determines a priori which subjects to teach as a function of the type of society that he considered preferable, Aristotle seems more flexible and discusses the central problem: what type of education should be preferred? There is no doubt, says Aristotle, that young people must learn subjects that are useful and indispensable, such as reading, writing, arithmetic, and geometry, and perhaps even drawing; but, later, which other subjects could be "useful"? In *Politics* (8.1) he notes a few simple and fundamental facts, unfortunately so often neglected today: the choice and the learning of a subject, for instance, vary according to personal and social orientation. Aristotle further asks what is more useful: to cultivate intelligence or moral character? Like Plato, he gives, at least in theory, preference to the second, that is, to the liberal arts. He states (1338 *a*), "It is evident, then, that there is a sort of education in which parents should train their sons, not as being useful or necessary, but because it is *liberal* or noble. Whether this is of one kind only, or of more than one, and if so, what they are, and how they are to be imparted, must hereafter be determined" (italics mine).

In synthesis: the two great Greek teachers, Plato and Aristotle, endeavored to ennoble the spirit and did this, inter alia and unlike Socrates, through unidirectional teaching; their teaching, however, had nothing to do with the vocational training modern higher education attempts to convey. And yet, ennobling the spirit could encourage new ideas with dubious or destabilizing results for society. Plato proposed selecting candidates for studies, with the counterproductive result of creating competition that could manipulate the goal of study, generate stress, and create an elite whose influence is questionable and may hinder progress (an elite member, e.g., an aristocrat, may actually not evidence so noble a soul after all, as instead he may become authoritarian and discriminatory). Aristotle, who was less elitist, laid the foundations to a division of fields that, today at least, may not always facilitate scientific study and may hinder progress. Their contribution to thinking and civilization remains, nevertheless, of fundamental importance.

Aristotle's teaching is a prelude to the grand philosophical and scientific flowering of the Hellenic world

Geographer and historian Strabo (c. 60 AD–24 BC), at the beginning of the Christian era, tells (in his *Geography* 13, 1.54) that "Aristotle bequeathed his own library to Theophrastus, to whom he also left his school; and he is the first man, so far as I know, to have collected books and to have taught the kings in Egypt how to arrange a library."[6] Aristotle's library and his teaching may have reached Egypt and inspired the founding of the highest cultural and scientific institution in antiquity, the Library of Alexandria and the annexed Museum (i.e., shrine of the muses). This intellectual and religious center promoted the written text as the basis for research and for collection and growth of knowledge.

The city of Alexandria, in Egypt, was founded between 332 and 331 BC by Alexander the Great. It soon became the fulcrum of various Mediterranean and Middle Eastern cultures: Greek, Egyptian, Hebrew, and the like. The documentation related to its Library and Museum, as the extensive study of Bibliophile Edward Parsons (1952) shows, is confused and at times legendary. The Library is said to have been founded at the beginning of the 3rd century BC by Ptolemy I Soter, founder of the Ptolemaic dynasty (r. 323–283 BC), under the direction of the Athenian orator, politician, and scholar Demetrius of Phaleron (c. 350–280 BC). The latter had studied at the Athenian Lyceum under Aristotle's successor, Theophrastus, and his presence attests the link between the Library and Aristotle's Lyceum. Other sources say the founder was Ptolemy I's son, Ptolemy II Philadelphus (r. 283–246 BC); in any case, under Ptolemy I, the related Museum was built to host an academic and religious community of scholars, inter alia, in the wake of the Egyptian tradition associating learning with the priestly class. The community was directed by a priest of the muses in analogy with the priests directing the Egyptian temples. Strabo recounts the following:

> The Museum is also a part of the royal palaces; it has a public walk, an Exedra with seats, and a large house, in which is the common mess-hall of the men of learning who share the Museum. This group of men not only hold property in common, but also have a priest in charge of the Museum, who formerly was appointed by the kings, but is now appointed by Caesar.[7]
>
> (Strabo, *Geography* 17. 9)

The Museum and Library of Alexandria can be considered the first important research institution in the Western world. It was, above all, a center of collection of written material: it is said that all scrolls on board of ships that docked in Alexandria's harbor were taken to the library and copied. The library kept the original, and the copy only was returned to the owner. As such it marked the beginning of the end of Western oral culture. A description (unfortunately unreliable) by 12th-century Byzantine poet and grammarian John Tzetzes (c. 1110–1180) reports that "In the outer library there were 42,800 volumes; in the palace library 400,000 mixed volumes and 90,000 single volumes and digests, according to Callimacus, a man of the court and royal librarian" (Parsons 1952, 110; Callimacus was a poet working under the patronage of Ptolemy II and Ptolemy III and "single volumes"— better said, rolls—contained only one work, or if the work was extensive, one part of it. They gradually replaced the longer "mixed" rolls.). At the time of Ptolemy III (r. 246–222 BC), there were probably two libraries: a larger one, located inside the royal palace, used by the Museum scholars, and a smaller one, for public consultation, located outside the royal court, in the temple of Serapides, or "Serapeum." The god Serapis combined Egyptian and Greek divinities in anthropomorphized form and was compatible with the culture of Ptolemaic Alexandria. The temple symbolized the symbiosis between the two cultures under the aegis of religion.

The Alexandrian complex then was more than a religious center for collection of material. The many works formed a vast nucleus of data allowing study and research in a range of fields such as literature, mathematics, astronomy, and more. Some historians have considered the complex as the prototype of the modern university. True or not, it appears at least that, like in modern universities, Alexandrian scholars were encouraged to carry out research and, perhaps, to teach *sine cura*; as classical philologist Rudolf Pfeiffer (1889–1979) jokingly remarks, "There was plenty of opportunity for quarrelling with each other" (Pfeiffer 1968, 97). One could better relate to a Renaissance or Early modern academy under court patronage, a subject treated later in this book. Pfeiffer adds,

It was also a revolutionary change that free public access was [given] to the immense written treasures of the Alexandrian libraries; they were not temple- or palace-libraries to which a privileged minority was admitted, but they were open to everyone who was able and willing to read and to learn. There was a free world of the spirit even in the new monarchies, and the preconditions for such a development existed only where Greek civilization prevailed. The unprecedented interest in books was kindled by the new scholar poets, who were in desperate need of texts; by a notable coincidence the royal patrons and their advisers immediately fulfilled these imperative demands in a princely way. We shall find a similar sequence of events when in the Italian renaissance the ardent zeal of the poets and humanists from Petrarch to Politian led to the recovery of the Classics and the setting-up of great libraries.

(Pfeiffer 1968, 103)

The above picture may be idealized, although the Museum owes, no doubt, much of its success to the Ptolemaic patronage, evidently in an attempt to strengthen its power by Hellenizing Egypt. The Ptolemaic dynasty, nevertheless, left room for free interchange between different cultures, and this may have been one of the secrets behind the extraordinary scientific results produced by the Museum.

The Alexandrian Museum contributed to a reevaluation of mathematics and manual and technical labor, even if only in the modern era—more precisely, from the 19th century on—did technology begin to achieve the dignity of a "pure" science. The Library of Alexandria welcomed scholars and scientists, among them Euclid, who, around 300 BC, systematized geometry; Aristarchus (c. 310–230 BC), a brilliant astronomer who, some 1,700 years before Copernicus, proposed that the Earth revolved around the Sun and may have inspired the latter; Archimedes (c. 287–212 BC), who may have stayed at the Museum in his youth (Schneider 1979, 7–8); and Eratosthenes (c. 276–194 BC), one of the librarians who, among others, by an ingenious procedure succeeded in calculating the circumference of the Earth. It is to Eratosthenes that Archimedes addressed a famous letter, discovered in the early 20th century, formulating a "Method" of geometrical calculus.[8] Hipparchus (c. 190–120 BC) succeeded in making most precise astronomical observations, leading him to discover the precession of the earth's equinoxes and to compile the first comprehensive star catalog of the Western world. Later, in the 2nd century AD, the most famous of all Alexandrian astronomers, Claudius Ptolemy (c. 90–168 AD), designed a complex, yet precise geocentric astronomical model that remained in use for almost fifteen centuries, until after Copernicus. Much research has been dedicated to these and other famous Museum's scholars, and yet, no in-depth research has been conducted (or could be conducted, due to the dearth of sources) regarding the function of the Alexandrian as a center of higher education. Indeed it must have been a center of research rather than teaching.

Medicine, too, developed in Ptolemaic Egypt, although probably outside the Museum, and medicine was one of the first fields of higher learning that freed itself from religion. Egypt had its own medical tradition based, inter alia, on magic, and the Greeks brought into the country a more rational approach originating mainly from the medical schools of Kos and Cnidus. Two Greek physicians, Herophilus (c. 335–280 BC) and Erasistratus (c. 304–250 BC), worked in Alexandria during the reigns of Ptolemy II and Ptolemy III and are considered the founders of the Alexandrian "school of anatomy." Herophilus is credited for the discovery of the nervous system and the distinction between different types of nerves, and Erasistratus for having mapped the blood vascular system. The latter came to Alexandria at old age after serving as royal physician to Seleucus I Nicator (c. 358–281 BC) of Syria. There is no evidence that either worked at the Museum; nevertheless, according to the Roman encyclopedist Aulus Cornelius Celsus (c. 25 BC–50 AD), they performed dissections of human cadavers and vivisections on prisoners made available by Egypt's rulers. We do not know how their "school of anatomy"—if such an institution existed at all—functioned, although we know Herophilus had apprentices who probably paid him. One of them was Philinus of Kos, considered the founder of the "empirical" medical school, or sect, mentioned by Galen (c. 129–199 AD; the great physician at the service of Marcus Aurelius and other Roman Emperors). Hellenistic medicine, in general, is characterized by organizations of physicians into doctrinal groups or sects, and students of medicine may have trained within these groups.[9] In Alexandria, like in other parts of the Hellenic world and the Roman Empire, there were rudimental schools of medicine, which consisted of physicians surrounded by a certain number of apprentices and associations of physicians who also taught the art. In the majority of cases, nevertheless, to study medicine meant entering a physician's family circle. So much for medicine in Alexandria.

For five centuries, Alexandria remained one of the greatest cultural and scientific center of the ancient world. The legend tells that, when Julius Caesar conquered the city, a fire destroyed a portion of the volumes conserved at the Library. There are different conjectures regarding the Library's demise. One of the factors was—as in the case of the Athenian schools—religious, and is related to the Christianization of the Roman Empire. In 380 AD, the emperor Theodosius I (r. 379–395) decreed Christianity the official religion of the Roman empire, and the ensuing tension led to the destruction of the Serapeum in 391 AD. Nonetheless, it appears that the Library of Alexandria survived until the Islamic era.

Hellenism originated a variety of schools

The galaxy of classical philosophy also saw the creation of many other schools, each of which had its own teaching methods. These schools have influenced, in one way or another, various aspects of modern higher education. An exhaustive catalogue would exceed the scope of this chapter. A brief treatment of two of them will suffice.

In Athens, the beginning of the Hellenic era saw the formation of the Epicurean school of philosophy, known as The Garden. The founder, Epicurus (341–270 BC), preferred the calm atmosphere of a garden to the frenzy of a gymnasium. Of interest are a number of his observations, quoted by Diogenes Laertius (Book 10, 121b), regarding scholars and how one should behave, both as a scholar and as a teacher:

> It is possible for one wise man to be wiser than another. The wise man will also, if he is in need, earn money, but only by his wisdom; he will appease an absolute ruler when occasion requires, and will humour him for the sake of correcting his habits; he will have a school, but not on such a system as to draw a crowd about him; he will also recite in a multitude, but that will be against his inclination; he will pronounce dogmas, and will express no doubts.[10]

Some of these affirmations speak for themselves and are indisputable. Some other affirmations are indicative: that a scholar could find the opportune moment to serve an absolute ruler indicates how few scholars and, implicitly, how few literate people there were at the time, in order to be in such demand. And yet other affirmations, according to present knowledge, would contradict each other: dogma, which Epicurus is said to have supported, is an obstacle for the advance of knowledge; all these topics are tied to the teachings that I will later discuss.

A second contemporary Athenian school, less dogmatic, just to mention in passing, was that of the stoics (*stoà* means "portico," the place where members of this school would meet). The central institution remained the Library of Alexandria, marking the apogee of Hellenic culture, which in turn interacted with neighboring cultures. It owed its success above all to Ptolemaic patronage, but perhaps also to the opportunity it gave to intellectuals of various provenance who took advantage of the easy access granted by the port to interact in relative freedom. Hellenic culture and science, in general, enjoyed a deep and vibrant exchange of views with the monotheistic religions and influenced their educational systems (see next chapter). In turn, these systems integrated and conserved part of the Library's patrimony after its demise.

Hellenic culture and education were adopted by the Romans, and higher education in Rome meant mostly tutoring in rhetoric. Roman learning took a direction that was more encyclopedic than innovative. This too, influenced the genesis of the future university studies. In the 1st century BC, Marcus Terentius Varro (116–27 AC), an erudite Latin author, composed the *Disciplinarum libri IX*: nine books, each one dedicated to one of the following arts: grammar, dialectic, rhetoric, geometry, arithmetic, astronomy, music, medicine, and architecture. The Romans, too, adopted the arts as the basis of the education of the *liber homo*, as referred in the 2nd century AD by Cicero. During the rule of Constantine the Great (between 306 and 337), Christianity became a dominant religion of the Roman Empire, and Christian religious education, initially so antagonistic towards liberal arts, was gradually introduced into higher learning. Indeed St. Augustine (354–430), a teacher of rhetoric before his conversion, did not reject the liberal arts outright: in his *De Doctrina Christiana* (*On Christian Instruction*), he suggested to use them as a tool to better understand the Scripture:

> All branches of heathen learning have not only false and superstitious fancies and heavy burdens of unnecessary toil, which every one of us, when going out under the leadership of Christ from the fellowship of the heathen, ought to abhor and avoid; but they contain also liberal instruction which is better adapted to the use of the truth, and some most excellent precepts of morality; and some truths in regard even to the worship of the One God are found among them.[11]
>
> (*De Doctrina Christiana* II, 40)

At a certain point the liberal arts were pared down to seven: the final two, which were more applicative, were omitted. Martianus Capella, a pagan author who lived in the 5th century AD, in his allegorical work *De nuptiis Mercurii et Philologiae* ("On the Marriage of Philology and Mercury"), divided them into two groups: *trivium* (grammar, dialectic, rhetoric; the term *trivium* is at the root of the word *trivial*) and *quadrivium* (geometry, arithmetic, astronomy, music). The work describes Philology's ascent into the sky, with the seven liberal arts, to marry Mercury, the god of eloquence and commerce; it is a sort of encyclopedia of classical erudition and became widespread in the High Middle Ages. The seven liberal arts became the basis of higher learning in the Middle Ages, which was created by combining Greek education with religious Christian education. There is, nevertheless, one other domain of learning related to universities: Roman Law Schools.

Another source of medieval higher learning were law schools of the Roman Empire

Roman law schools have both religious and corporate (to use a modern term) roots. They can be traced back to the College of Pontiffs, a body that counseled the royal and early republican Roman rulers (in Latin *Collegium Pontificum*: from *co*, "together," and *-legia*, "law"; *Pontifices* is derived from *pontem facere*, i.e., "bridge builders"; originally the role of pontiffs may have been to placate the gods and spirits associated with the Tiber River). Initially pontifical learning was kept secret. Then in 254 BC, Tiberius

Coruncanius (d. 241 BC), Roman consul and the earliest plebeian *Pontifex Maximus* (high priest), apparently offered himself publically as a jurisconsult (i.e., as an expert qualified to advise on legal matters—in modern terms: a lawyer), initiating an era of public legal instruction in which a jurisconsult could collect students and teach law freely. At the time of Augustus (63 BC–14 AD) two rival law schools sprang up: the more traditional "Proculians," founded by Marcus Antistius Labeo (d. c. 10 AD), and the progressive "Sabinians," founded by Gaius Ateius Capito (c. 30 BC–22 AD). Both founders

> Gave legal instruction after the traditional fashion prevalent under the republic, which was to allow young men to be present as listeners while the jurisconsults gave opinions and to permit them to see how they conducted their law business, occasionally arguing with their pupils but rarely giving private instruction by means of connected lectures. This practical instruction was, under the empire, supplemented by teaching the students the elements of law.
>
> (Sherman 1908, 501)

According to 2nd-century jurist Sextus Pomponius, Sabinus charged for his teaching and was probably the first to found a law school proper. The Proculians eventually died out, and the Sabinians took over the instruction. Toward the end of the 2nd century, law schools spread out in Rome and in the empire; the best known was established at Berytus (today's Beirut). Under the Emperor Diocletian (r. 284–305) law schools were recognized by the state: Berytus first, and the schools of Constantinople, Alexandria, and Caesarea later. In 425 Emperor Theodosius II (r. 402–450) founded in Constantinople a center of teaching, known as "Pandidakterion" and sometimes referred to as a "university," for law, philosophy, the arts, Greek, Latin, and other subjects. In 533, however, Emperor Justinian suppressed the schools of Caesarea and Alexandria, but not the Berytus school, which contributed to the Codex of Justinian, the collection of fundamental works in jurisprudence, issued from 529 to 534.

> Under the later empire, systematic legal education was established in the course of time, and law students were obliged to study prescribed books in a certain order. [. . .] Furthermore, candidates for admission to the bar were required to pass a public examination before gaining the privilege to practice law.
>
> The required course of law study consisted of three years of study spent under professional instruction and two additional years of private study, thus altogether amounting to a five years' course. Justinian's reforms consisted largely in prescribing that thereafter—533 A.D.— students should be taught exclusively from the Justinian law books: for the books hitherto used by students did not suffice to give satisfactory

a legal equipment in the sixth century, owing to the condition of the Roman law prior to its complete codification under Justinian.

(Sherman 1908, 505)

When, in 554, Justinian reconquered Italy, he applied there the method of teaching in the Eastern Empire, a tradition that was somehow preserved and reemerged in Bologna, where the first medieval university was founded, initially as a law school (see chap. 6).

CONCLUDING REMARKS

The classical educational institutions already displayed some of the conflicts that can be found in the modern university: on the one hand, spiritual, intellectual, and scientific development with all the relative social benefits; on the other hand, dogmatism, careerism, elitism, and also stress intentionally induced by extrinsically motivated study, which was to become so typical of the modern university. Note, also, the strong link between higher learning and religious worship, a link that will become even more evident in the next chapter.

BIBLIOGRAPHICAL NOTES

Besides Diogenes Laertius, and the modern authors Ross, Cherniss, and the informative Lynch, all mentioned in the preceding chapter, Parsons's *The Alexandrian Library* (1952) is an exhaustive study of the sources regarding the Library of Alexandria. A more recent scholarly and well-written Italian work is *La biblioteca di Alessandria: Storia di un paradiso perduto* (2010), by Monica Berti and Virgilio Costa. *The Library of Alexandria: Center of Learning in the Ancient World*, edited by Roy MacLeod (2010), is an excellent collection of articles on specific aspects of the Library of Alexandria. An overview of Roman law schools is proffered by Charles Sherman's "Study of Law in Roman Law Schools" (1908).

NOTES

1. http://classicpersuasion.org/pw/diogenes/dlaristotle.htm, accessed January 1, 2012. Diogenes Laertius puts Aristotle's return to Athens before his migrations.
2. http://classics.mit.edu/Aristotle/sophist_refut.mb.txt, accessed January 10, 2012.
3. http://classics.mit.edu/Aristotle/posterior.1.i.html, accessed July 18, 2012.
4. Aristotle also had a theory of less proper learning, such as generalization, which is not really knowledge as it does not entail certainty.
5. http://classics.mit.edu/Aristotle/politics.8.eight.html, accessed January 1, 2012.

6. http://penelope.uchicago.edu/Thayer/E/Roman/Texts/Strabo/13A3*.html, accessed December 4, 2012.

7. http://penelope.uchicago.edu/Thayer/E/Roman/Texts/Strabo/17A1*.html, accessed December 5, 2012.

8. The existence of Archimedes's "method" was already known during the Renaissance, but no trace remained. Only in 1906 did the science historian Johan L. Heiberg (1854–1928) discover the manuscript in a library in Constantinople; see Heath's *The Method of Archimedes* (1912). It was one of the most sensational discoveries in the modern history of science.

9. I rely on Vallance's (2010) excellent article emphasizing the tension in Alexandria between physicians relying on written treatises and empirical physicians, and on von Staden (1989)—who conducted a thorough study on Herophilos together with a collection of related sources.

10. http://www.attalus.org/old/diogenes10c.html, accessed January 1, 2012.

11. http://www.intratext.com/IXT/ENG0137/_P2B.HTM, accessed April 11, 2013.

4   The Religious Roots
Priests and Rabbis

In ancient Judaism, learning was part and parcel of religion

Knowledge, in general, requires legitimization, and the highest legitimization is divine. As seen in the previous chapters, Mesopotamian, Egyptian, Greek, and Hellenic higher learning was strictly connected to religious worships and a monopoly of priestly castes (Greek philosophers, however, began to mark an exception). The same occurred in Jewish monotheism, persisted, as we shall see, in the Islamic and Christian world, penetrated deeply into the European university, and has, to a certain extent, left traces even today. In a number of modern Western universities on the cutting edge of science, the faculty of theology formally maintains a place of pride.

Every religion has its own incontestable doctrinal nucleus. The basis of the monotheistic cult was set forth by Moses, presumably in the 13th century BC. Moses's laws, regulating religious as well as daily life, are included in the Torah (which means "teaching" or "instruction"), or Pentateuch—the first five books of the Old Testament. A great deal of related rules were, however, transmitted, discussed, and developed orally, forming the *Torah shebe'alpe*—the Oral Torah. Herein, too, lies a general socio-political aspect. Because, in antiquity even more than today, the written word was more authoritative than the spoken one, writing down what was debated might have led to questioning the Torah and the authorities whose duty was to put its teachings into practice.

The Old Testament remains the primary source to our knowledge of ancient Judaism, but beyond a few hints, it offers little historical information regarding the teaching and learning methods and even less about possible educational institutions. In general, Jewish education fulfills a precept that comes a few paragraphs after the Ten Commandments: "You shall teach them [the laws] diligently to your children, and shall talk of them when you sit in your house, and when you walk by the way, and when you lie down, and when you rise" (Deut. 6: 7). This precept is so important that religious Jews recite it, as prescribed, twice a day, during morning and evening prayers. It makes education, or at least indoctrination, mandatory, at least in theory and on an elementary level, and indicates that it was initially transmitted from father to son, as common in antiquity. As far as this Torah precept is concerned, one could say that education and religion are one.

Beyond the Old Testament, around the end of the 2nd century AD the Oral Torah was partially written down in the *Mishnah* (a Hebrew word with several meanings: "learning" [by heart], "repeating," and also "secondary" [to the *Torah*]). The *Mishnah*, in turn, was subjected to further debate, which, around 500 AD, was collected in the *Gemara* (from Aramaic "[to] study;" the *Gemara* is written mainly in Aramaic). The *Mishnah* and the *Gemara* together form the *Talmud* (meaning "learning" or "teaching"). There are two *Talmuds*: one "Babylonian," the other "Jerosolimitan" (*Yerushalmi*) or "Palestinian." Today one normally uses the term *Talmud* for Babylonian *Talmud* and even for its *Gemara* alone. These treatises, if taken in an appropriate context, can offer significant historical information concerning ancient Jewish learning.

How literate were ancient Jews? Historically, they are considered the most literate of all the ancient populations. During the Renaissance and later, it was believed that ancient Hebrew knowledge was, along with classical knowledge, the most luminous example of past civilizations, and their learning influenced the university and the modern higher learning considerably. It is therefore amazing that their higher learning has been neglected by most literature dealing with history of higher learning (e.g., Clarke 1971). To be sure, it is hard to say if at all, when, and to what extent the ancient Jews became more literate than the Gentiles. We lack documentation, particularly concerning the period of the First Jerusalem Temple, built, according to the Old Testament, under King Salomon in the 10th century BC and destroyed by the Babylonians in 586 BC. During this early period, and much later as well, Israelites were still an oral culture and were as literate as the bordering Egyptian and Mesopotamian subjects, that is, a very small percentage of them could read and write. It took them many centuries to become literate, and all sources clearly document the interdependence between literacy and religion.

Religion, learning, and social stability are strictly related

According to Durkheim, one of the founders of modern sociology, religion is a "social fact"—implicitly a factor of social cohesion (Durkheim 1982). Durkheim, whose life straddled the 19th and 20th centuries, descended from a family of rabbis from Lorraine. After becoming agnostic, he embarked on a university career in the field of education, and in 1896, in Bordeaux, he was assigned the world's first chair of sociology. His holistic orientation in sociology might be debatable, but his erudition and knowledge of both Judaism and the contemporary theories of education is indisputable.

In Durkheim's viewpoint, which is still fairly present in sociology and influences other fields, Hebrew religious education could be considered a function of social stability, a means for keeping the population united, or at least the tribes that formed it.

Religion, education, and social stability are strictly related. In the Greek Polis each city had its patronal divinity, itself a means of cohesion. Greek higher education was related to religion and had a social function of central importance; this is not the only element it has in common with Hebrew education. Hebrew education, however, was strictly religious, inter alia, because religion embraced all aspects of life. Greek formative learning had more specific goals: for example, sophist and Socratic education aimed at cultivating virtues; Platonic education aimed at creating an elite. Hebrew education was conceived for a people who considered themselves in many ways, *in toto*, an aristocracy, and took universal obligations upon themselves; in theory it should not have had an elitist function in the commonly held form, with more rights, and claims to rights, than obligations. Above all, it aimed to prepare its believer for worship. This does not mean that the ancient Israelites did not have their own elites, as the Jewish historian Flavius Josephus (c. 37–100 AD) remarks at the beginning of his autobiography, written at the end of the 1st century AD:

> As nobility among several people is of a different origin, so with us to be of the sacerdotal dignity, is an indication of the splendor of a family. Now, I am not only sprung from a sacerdotal family in general, but from the first of the twenty-four courses; and as among us there is not only a considerable difference between one family of each course and another, I am of the chief family of that first course also.[1]

Born Joseph Ben Matitiahu into a sacerdotal family, Josephus had been the military governor of Galilee during the great rebellion of the Jews against the Romans, which erupted in 66 AD and ended with the destruction of the Second Temple by Titus (39–81 AD) in 70 AD. He gave himself up to the Romans, hitched his destiny to Emperor Vespasian (r. 69–79 AD), changed his name to Flavius Josephus, and spent the rest of his life in Rome writing history books about the Jews, which, above and beyond their clarity, are a highly important source of information on contemporary Judea. His autobiography might be apologetic because he was accused of having betrayed his people, yet it offers some hints as far as Jewish ancient education is concerned:

> I made mighty proficiency in the improvements of my learning, and appeared to have both a great memory and understanding. Moreover, when I was a child, and about fourteen years of age, I was commended by all for the love I had to learning; on which account the high priests and principal men of the city came then frequently to me together, in order to know my opinion about the accurate understanding of points of the law.[2]

The passage hints that studying was partly—if not entirely—mnemonic, typical of an oral culture.

The sacerdotal caste, to which Josephus belonged, fulfilled the onerous and complex task of both administering the cult and safeguarding knowledge. Like in the European aristocracy, one belonged to it—and continues to belong to it, even if with few symbolic functions—by birth. Hebrew priesthood was (and remains) formed by the tribe of the Levites—the descendants of Levi, one of the twelve sons of the patriarch Jacob, to which Moses himself belonged. In the past it enjoyed many privileges but also had burdensome obligations to fulfill, and in many aspects its members were disadvantaged. For example, as opposed to the other eleven tribes, no territory was assigned to the Levites. It was, and formally still is, stratified in major and minor priesthoods.

The major priesthood, the *cohanim*, was composed of the descendants of Aaron, Moses's brother and the first High Priest; their duties involved mainly, but not only, in offering sacrifices and in priestly blessing. One additional duty of theirs was ancestral to one of the first medieval university teachings: medicine. The Leviticus, the third book of the Old Testament, dealing with priestly duties, devotes four chapters (12 to 15) to topics such as child birth, leprosy, or bodily secretions. The treatment follows considerations of purity and impurity and is oriented primarily at protecting the community, and the priest had ultimately a shamanic role. From the sociological point of view it displays a typical tribal performance tending to holism. It contains, however, interesting elements as far as the history of medicine is concerned (as does the sacerdotal function in all primitive societies): the care for communal hygiene and the seclusion of infected individuals. Such places of seclusion later became centers of interest for scholars and physicians who studied the illnesses and could be considered the seed of a hospital and a medical school. One of the first universities, the medical school of Salerno, started as a mere collection of physicians in a health resort. The *cohanim* could eventually also act as judges. All these duties of the *cohanim* required a particular learning, and the historian André Lemaire (1981, 68–70), who studied the thin evidence offered by ancient Hebrew inscriptions and texts, conjectures that there must have existed a higher level school that taught priests, at least, the complex rules and practices of sacrifices.

The minor priesthood was composed of the rest of the Levites and is generally referred to as "the Levites." According to the Old Testament's Second Book of Chronicles (34: 12–13), "The Levites—all who were skilled in playing musical instruments—had charge of the laborers and supervised all the workers from job to job. Some of the Levites were secretaries [*shotrim*], scribes [*sofrim*] and gatekeepers [of the Temple]."[3]

The term *shotrim* is derived from an Akkadian root referring to writing (in modern Hebrew the term is used for policemen). The *shotrim* are mentioned several times in the Old Testament in company of high officials, such as Judges or elders. David's and Salomon's kingdoms (11th–10th centuries BC) and the two kingdoms that split from the latter (Israel in the north and

Judah, with Jerusalem, in the south) undoubtedly needed bureaucrats and scribes. The same can be said concerning a Temple of the size of Salomon's. Lemaire (1981, 67–68, 70–71) further conjectures the existence of a royal school for civil servants and a school for prophets. "Prophet" (in Hebrew *navi*, originally meaning spokesperson), in ancient Israel, could have had a meaning broader than that of an intermediary between the divine and humanity: it could refer, more generally, to a learned guide. At least some of them, such as Jeremiah and Ezekiel, were priests (*cohanim*). In any case, prophets, like civil servants, must have had a broad knowledge, yet there is no direct evidence of the existence of schools for them. Most probably the required wisdom was transmitted from father to son and the method of study was mnemonic, perhaps using proverbs and chanting as didactical means.

The role of scribes became increasingly relevant during the Second Temple (535 BC–70 AD). The figure Ezra *hasofer* (Ezra "the scribe"), to whom a book of the Old Testament is devoted, is paradigmatic. Ezra was a priest—perhaps even the high priest. The king of Persia had sent him to Judea—then a province of Persia—to reintroduce the law of God in Jerusalem, inaugurating the era of the Second Temple. He can be taken as the watershed between the priest's and prophet's era of leadership during the First Temple, and that of the scribes (not necessarily *cohanim*) during the Second Temple. The Levites, then, came to fulfill also the duties of secretaries and scribes. Moreover, the book of Nehemiah—the Jewish governor Judea who collaborated with Ezra—reports the following impressive mass schooling event, in which the Levites assisted Ezra in his mission:

> all the people came together as one in the square before the Water Gate. They told Ezra the teacher of the Law to bring out the Book of the Law of Moses, which the Lord had commanded for Israel. So on the first day of the seventh month Ezra the priest brought the Law before the assembly, which was made up of men and women and all who were able to understand. He read it aloud from daybreak till noon as he faced the square before the Water Gate in the presence of the men, women and others who could understand. And all the people listened attentively to the Book of the Law. Ezra the teacher of the Law stood on a high wooden platform built for the occasion . . . Ezra opened the book. All the people could see him because he was standing above them; and as he opened it, the people all stood up. Ezra praised the Lord, the great God; and all the people lifted their hands and responded, "Amen! Amen!" Then they bowed down and worshiped the Lord with their faces to the ground. The Levites [. . .] instructed the people in the Law while the people were standing there. They read from the Book of the Law of God, making it clear[a] and giving the meaning so that the people understood what was being read.[4]
>
> (Neh. 8, 1–7)

The Levites, on this occasion, acted as lectors and probably even as teachers to a gathering of illiterate people. More in general, they became part of the new intellectual leadership of the people of Israel. This brought them in conflict with the *cohanim*, who gradually lost their functions and were demoted to Temple rituals. To be sure, the highest authority in Israel during the Second Temple period seems to have been the "Great Assembly" (*Knesset hagedolah*) of 120 sages, including scribes; the available documentation does not say when it was founded and what was its exact role, and some historians doubt it ever existed. The link between higher learning and religion is, however, evident.

Later, at the end of the 4th century BC, Judea became part of the Hellenic world, first under Ptolemaic Egypt and later under the Seleucid Empire. Greek cities were founded in Palestine with gymnasiums and schools introducing Hellenic polytheistic culture into the country. The two cultures interacted initially with curiosity, interest, and even enthusiasm. The story of the Septuagint, which tells how seventy-two Jewish scholars were asked by the Ptolemy II Philadelphus to translate the Torah into Greek for the Alexandrian Library, although legendary, is emblematic. Hellenism spread in Judea, at least among wealthy Jews and young aristocrats, creating a new class of Hellenized Jewish literati. Tensions were nevertheless bound to emerge. The First Book of Maccabees (1: 14–15)—a deuterocanonical book, that is, included in the Christian Old Testament, but not in the Hebrew Bible— complains: "They built a place of exercise at Jerusalem according to the customs of the heathen. And made themselves uncircumcised, and forsook the holy covenant, and joined themselves to the heathen, and were sold to do mischief."[5] The result was a revolt (167–160 BC) led by the Maccabees—a Judean rebel group—against the Seleucid Empire. There followed a twenty-five-year period of Jewish independence till Judea came under Roman influence. It is in this period that other cultured groups emerged in Judea, the most renowned being the Sadducees and the Pharisees.

There are many conjectures concerning the origins and the nature of these groups. The Sadducees must have been a relatively small group, active from the 2nd century BC through the destruction of the Second Temple, and formed or belonged to the upper, social, and economic class. The origin of their name is not quite clear, except as it is related to the Hebrew word *zedek*—meaning "justice." They were probably priests, or were related to priesthood, and fulfilled various political, social, and religious roles, including maintaining the Temple. They seem to have favored Greco-Roman culture, but were conservative in religious matters. They regarded the written Torah as the sole source of divine authority and belittled the importance of the Oral Torah.

The need of following the orally transmitted Torah and exegesis was nevertheless stressed by another prominent group, the Pharisees. They must have emerged, at least partly, out of those scribes and sages who harked back to Ezra and, possibly, the Great Assembly. Their name (in Hebrew

Prushim—"separatists") could be a result of their self-imposed separation from the gentiles and probably also from less observant Jews. They insisted on the biding of the orally transmitted teaching, in addition to the written teaching, and developed a Hermeneutic. In his *The Wars of the Jews* (2, 162), Josephus writes, "The Pharisees are those who are esteemed most skillful in the exact explication of their laws."[6] Julius Wellhausen (1844–1918), the (protestant) pioneer of critical biblical studies, conjectures that both were guild-like groups (Wellhausen 2001, 8). The Sadducees was the guild of priests—the guardian of the laws—struggling to keep its power and privileges, whereas the Pharisees was the guild of scribes that ultimately, after the destruction of the temple, replaced the Sadducees as guardian of the law.

All the members of the above mentioned groups, which could eventually intersect, as well as other learned authorities mentioned by the various texts, received advanced schooling in one way or another, although we do not know exactly how. As Catherine Hezser's (2001) monumental work of Jewish literacy in Roman Palestine points out, contemporary Hebrew education, whether elementary or advanced, was a complex matter. The earliest Hebrew text to mention a school proper is the Book of Ben Sira (also known as *The Book Ecclesiasticus*, or *The Book of the All-Virtuous Wisdom*, 51, 23). Its author, Shimon Ben Sira (also known as Jesus Ben Sirach, or Siracides), was a Jew from Jerusalem, and the Book may have been written—or rather compiled—at the end of the 3rd or at the beginning of the 2nd century BC in Alexandria, where Ben Sira is thought to have established a school. His *Ecclesiasticus* is, like the First Book of Maccabees, a Deuterocanonical book, and the Hebrew term used for school by Ben Sira is *beth midrash*, meaning "house of interpretation" or "house of exegesis," although we do not know exactly what it meant in this case.

What wisdom was taught under the Second Temple and how? Without a doubt, the Torah, learnt most probably by heart. An elementary education was necessary in order to follow the dialectic of the *midrash*: students had to at least be able to read (not necessarily write). Towns may have occasionally had elementary schools: the Babylonian *Talmud* reports that high priest Yehoshua Ben Gamla, shortly before the destruction of the Second Temple, "ordained that teachers of young children should be appointed in each district and each town and that children should enter school at the age of six or seven" (Tractate *Baba Bathra* 21a—meaning "the last gate" in Aramaic, i.e., the last of the three tractates dealing with responsibilities and rights of property owners).[7] At a higher level came the Oral Torah and the exegesis, which might have formed the basis for debate and thus made study, inter alia, more interesting. The teachers were erudite masters of law called rabbis (from *rav*, literally "great," to indicate their great knowledge and prestige). Rabbis are not priests, even though nothing prohibits them from performing the two functions. They were sages who replaced priests in safeguarding knowledge. Gradually the monopoly of the cast of clerics over knowledge was broken and replaced by another type of clerical, or pseudo clerical, more open group. As long as the Temple stood, the task of the *cohanim* was restricted to administering the cult.

Jewish higher learning consisted mainly of studying the law

How did a student become a Rabbi? He most probably had to attend a *beth midrash*, or at least spend a few years as a rabbi's apprentice. The way the *batei midrash* (pl. of *beth midrash*) functioned remains—despite the quantity of available exegetic writings—a purely conjectural matter. They were normally attached to synagogues—local houses of prayer—and most probably informal. Presumably, rabbis discussed the application of the laws with other rabbis or with their students—exclusively males—who most likely were seated around them (the traditional exclusion or discrimination of women in the orthodox Jewish educational system persisted into the modern era, despite the progress made by higher learning institutions in the last centuries.). The debate was honed through a dialectic of questions-and-answers and critical observations.

A student that had achieved an adequate level of proficiency could be officially ordained, and here we have a first step towards an academic title. "The academic degree was originally unknown to Druids and Huns, Romans and Greeks, Arabs and Zulus, Israelites and Philistines, Persians and Chaldean, Hindus and Chinese," argues historian of science William Clark in a mocking article "On the Ironic Specimen of the Doctor of Philosophy" (1992, 97; 2006, 183). Clark may have not been totally accurate. The Hebrew term for ordaining is *semikha* meaning literally "leaning" or "laying" [of hands]; the term also means to "rely on" or "to be authorized." Moses is told to have ordained Joshua—his follower as leader of the Israelites—by laying of his hands upon him (Num. 27: 18–24 and Deut. 34: 9). According to the Babylonian *Talmud* (*Sanhedrin* 13*b*) the ordination of a rabbi did not require the formal laying of the hands but had to be conferred by three sages, of whom at least one had to be ordained as a rabbi. By declaring the candidate rabbi they granted him the authority to teach, pronounce sentences related to the application of the law, or plainly act as a judge, albeit with a series of limitations in place, time and legislative domains (such as those of laws of kashrut or monetary issues). The *Talmud*'s *Sanhedrin* tractate discusses these limitations at length. The *semikha* and its limitations were aimed, much like in Islam later and unlike European universities, to ensure a rigorous application of the law. There is no direct evidence that rabbis gained money from their teaching and jurisprudence; if not well-off, some made their living from other sources or professions.

In Judea, as well as in other parts of the Jewish diaspora—above all in Babylonia—, studies of the (religious) law were also closely related to judiciary practice. The Judean judicial system during the Second Temple, like the educational one, was complex. A Jewish informal and basically religious judicial system was active in parallel to the Roman civil courts. The related documentation, however, is scattered and confused: it mentions a supreme legislative body, occasionally called "Great Sanhedrin" (from the Greek word "*synedrion*," meaning "sitting together," hence "assembly" or "council"—the term is an indication of the strong Greek influence), with 70 or 71 members, who met in the Chamber of Hewn Stone in the Temple outer, least sacred, Courtyard in Jerusalem. There were then lower courts, "Lesser Sanhedrin," with twenty-three members presumably acting all over the country. Only ordained rabbis could sit in the Sanhedrin (*Sanhedrin* 13*b*; Steinsaltz 1996, 138). Single rabbis as interpreters of the law, too, had faculties of judging in smaller rabbinical courts, *batei din* (houses of judgments, pl. of *beth din*), with one, three, or perhaps more rabbis. The *beth midrash* became occasionally *a beth din*. During debates in these centers, whether scholarly or legal, participants could gain knowledge of the laws. The *beth midrash* can therefore be considered an informal law school. The University of Bologna, deemed to be the first University in the Western world, too, started, to repeat, in the Middle Ages as a law school. In Judaism, as later in Islam and in early Christianity, religious law regulated all aspects of life, and it is natural that it became the first subject of higher learning.

Religious law in late antiquity was still far from dealing with domains that one would today in science. However the exegesis could eventually deal with related topics. An interesting example is that of the application of theoretical astronomy to the fixing of high holidays. The Hebrew calendar is both solar and lunar, yet twelve lunar months add up to less than one year so that leap years have thirteen months each. During the Second Temple the beginning of a new month was determined by the observation of a new moon: according to the *Mishnah* (*Rosh Hashana* 2), at least two witnesses had to attest to the Temple's authorities in Jerusalem having seen the new moon. The Temple declared the beginning of the new month and spread the news throughout the Jewish world. After the destruction of the Second Temple this procedure could no longer carry on. The religious authority was at that time partly resumed by some leading *batei midrash*, such as the one active in the city of Yavne, in southern Judea, and later in Usha, in Galilee. The Yavne *beth midrash* was also known as "the vineyard," probably because its students used to sit in rows that called to mind rows of grapevines. Yavne and Usha, then, were more than just schools. At the end of the 1st century AD, Raban Gamaliel II ("Raban" was a title given to the head of the Sanhedrin), who presided both institutions, introduced several reforms, including the replacement of traditional, empirical procedures with calendar computations based on Greek astronomy—an early example of where science supports religion and religious exegesis adapts to science. Galileo later depicts this kind of collaboration in his *Letter to the Grand Duchess*

Christina (1615), a short brilliant treatise concerning the relation between religion and science, with the following words: "The Holy Scripture and nature derive equally from the Godhead, the former as the dictation of the Holy Spirit and the latter as the most obedient executrix of God's orders" (Finocchiaro 1989, 93).

Back to Jewish learning, after the fall of the Second Temple and the dispersion of the Jews, and even more after the defeat of the last major rebellion against the Romans under Hadrian (re. 117–138 AD), in the years 132–136 AD—known as the "Bar Kokhba revolt," from the name of its leader—Judaism underwent a religious and intellectual decentralization, both on the geographical and the institutional level. The institutional decentralization is a corollary of the geographical one and is closely tied to education. It is probably at that time that Jews started becoming more literate. The *batei midrash* grew in importance and substituted the Temple and its sacrifices with prayer in a new era of worship that was more intellectual than practical. So did the status of rabbis, to the point that the Romans under Hadrian, well aware of the importance of rabbinical ordination, decreed "that whoever ordain will be killed, and whoever is ordained will be killed, and the city where they ordains will be destroyed, and the district in which they ordain will be uprooted" (*Sanhedrin* 14a; Steinsaltz 1996, 139). The practice was carried on illegally, but with time and with the thinning out of the Jewish population in Palestine, it stopped.

The *batei midrash*, nevertheless, spread throughout the Jewish diaspora. Initially they did not seem to have any particular canonical form; much the same occurred later in Islamic higher learning institutions and European universities and Renaissance learned societies. Judaism, like Christianity and Islam, is not monolithic, and the many currents and sects had their own traditions and religious and educational institutions.

The *beth midrash*, as an institution, developed into the *yeshiva* (literally "sitting")—a school of intermediate to higher religious learning that still exists today, and Yavne and Usha may, perhaps, be considered seeds of *yeshivot* (pl. of *yeshiva*). In Jewish tradition the *yeshiva* distinguishes itself from the *beth midrash* for being a more permanent and more formal learning institution in which students study for a longer period under the supervision of rabbis. The wisdom that *yeshivot* produced bestowed on them, rightly or wrongly, the connotation of "academies" to underline their similarity to analogous Greek institutions as well as to modern ones. The *Mishnah* and *Gemara* helped maintain the cohesion of the autonomous Jewish communities as one people and were complemented by a vast Rabbinic literature discussing and interpreting the laws. The use of titles was spread despite the lack of official religious ordination: the sages of the *Mishnah* got the honorific title "Rabbi," and the sages of the Babylonian *Talmud*, "Rav."

As Shelomo Dov Goitein (1900–1985), the scholar who last century studied the "Cairo Geniza"—an impressive collection of manuscript fragments found in the *geniza* ("storeroom") of a synagogue in Fustat (Old Cairo), from the 9th to the 19th centuries—, remarks, the first *yeshivot* united the functions of an academy, a parliament, and a supreme court. The *Yeshiva* in Palestine was called *Chavura* (which could be translated as "social group," or "institution"), or *Chavurat ha'tzedek* (the "righteous group"), and consisted of "scholars qualified to interpret and to administer the sacred law and themselves living according to its strictest standards" (Goitein 1971, 196), that is, a guild-like institution. The European universities, too, were initially guilds teaching, inter alia, the law.

Jewish ancient higher learning applied a variety of methods

The Mishnaic tractate of *Avot* ("fathers") contains ethical teachings "of the fathers," including maxims related to teaching and education, among them: "Five years is the age for the study of Scripture. Ten, for the study of Mishnah [. . .] Fifteen, for the study of Talmud"[8] (*Avot* 5, 22). According to Gerald Marvin Phillips (1956) and Steinsaltz (1989, 17–18), there were three levels of learning: primary, middle, and higher. In middle learning, the *Mishnah* was discussed, whereas in higher learning, which even Phillips calls "academy," the *Gemara* was studied. To me, this division seems excessively schematic and a posteriori. It must be remembered that the *Gemara* was only completed around 500 AD. What was studied at the Hebrew "academies" before that time? Moreover, the study of the *Mishnah* may require no less preparation than the study of the *Gemara*. The *Talmud* is the result of an oral tradition, and in such a context it is difficult to determine what came before and what came after.

The *Talmud* is scattered with remarks related to learning, at times contradictory.[9] Some, reminiscent of Plato, urge coercion, such as (*Baba Bathra* 21a): "stuff them with Torah like an ox . . . When you punish a pupil, only hit him with a shoe latchet."[10] The *Talmud* also encourages competition between students: "The jealousy of scribes increaseth wisdom."[11] Other remarks encourage mildness and tolerance, such as, "The dignity of your student should be as precious to you as your own" (*Avot* 4, 12).[12] Coercion is natural in a context of religious precepts, especially if applied to youths. In general one could say that the Talmud does not exclude it but recommends applying it in a mild, judicious way. One maxim that should be written in large letters above the entrance to every school and university:

The pedant does not teach and the timid does not learn.[13]

(*Avot* 2.6)

(The maxim could perhaps be improved to: "The pedant does not teach *properly* and the timid does not learn *properly*.") One of the aims of the present book is to point out and explain the importance of this maxim.

The pedantry of the paradigmatic professor, just like the timidity of the paradigmatic student, is not in the best interest of studies but is still present today, especially in universities, a result of the pervasive authoritarianism in Western education and of the erroneous custom of considering science, having after all supplanted religion, authoritative and thus a perfect field, conveyor of absolute truth.

Ancient Hebrew dialectic, which was studied in detail in its various forms, inter alia, by Phillips, was a part of the *dyun* (debate) conducted at the *batei midrash* and the *yeshivot*. To discuss its various aspects would go beyond the scope of this chapter;[14] let me only remark that Hebrew dialectic is not of Socratic type, that is, not open to any opinion and criticism, nor is it dedicated to developing individual virtues. It is more similar to Aristotelian dialectic because it limits itself to discussing the application of absolute principles, and indeed it might have been influenced by Greek thought. Nevertheless, as the 14th-century Spanish Jewish philosopher Hasdai Crescas (c. 1340–1410) noted in his work *Or Adonai* (Hebrew: "Light of the Lord"), a first, important medieval critique of Aristotelian physics and cosmology that heralded the gradual abandonment of Aristotelian teaching. Jewish principles, Crescas mantains, dealing with divinity and not with nature, are very different from the Aristotelian. According to Harry A. Wolfson (1887–1974), who translated the first book of this work into English (1929), Crescas questions Aristotelian dogmas, raising, as early as in the Middle Ages, the possibility of incompatibility between science and dogma.

Studying in *batei midrash* and *yeshivot*, unlike medieval universities, had relatively little utilitarian ends. Yet, as in future institutions of higher learning in the West (universities and, to an even greater extent, technical colleges), the Jewish teachers, if not rich themselves, made their livelihood through other activities, which could even be humble and manual. For example, Johanan the Alexandrian, a rabbi who lived in the 3rd century AD in Alexandria and whose teachings are quoted in the *Mishnah*, was also called Johanan *Hasandlar* ("the cobbler"), implying he earned his living as a shoemaker; Isaac *Nappaha* (Isaac the smith) who lived in the second half of the same century in Galilee was, as his title indicates, a smith. One of the fundamental precepts of Judaism is the observation of the Sabbath, when every form of manual labor (*melacha*) is strictly forbidden. This implies, inter alia, no less than six days of work, including manual work, which—always in respect of the precepts of the Torah—does not seem to have been limited by any moral reserve. Moreover, some precepts even called for specific manual training. In particular, the *Shechita* (butchering as prescribed by the Bible, which, for those times, minimized the suffering of the animals), transcription of the sacred texts—performed by the *sopher*—, and circumcisions. As opposed to the Greeks, who looked down on manual labor, these procedures were not carried out by slaves but by pious skilled manual workers with proper knowledge of the laws. It is not clear how widespread was the traditional reluctance to make religious learning a means of livelihood, and evidence goes both ways. Certainly then, as today (fortunately much less), those who had means had better access to studies and knowledge.

The Babylonian Yeshivot introduced formalism in learning

In the 3rd century, at least two famous *yeshivot* or academies were founded on the Euphrates and remained active for many centuries.[15] One, in the south of the country west of the River, was in the town of Sura, and the other in Pumbedita, close to the modern-day city of Fallujah. Learning was initially most probably informal (Hezser 1997, 195–196). After the compilation of the *Talmud*, the heads of these higher learning institutions were for centuries considered the spiritual leaders of the Jewish diaspora and got the honorific title *gaon* (meaning "pride" or "splendour" in Biblical Hebrew and "genius" in modern Hebrew; pl. *geonim*). By then the *semikha*, which was generally limited to the land of Israel, was no longer practiced, but the *geonim* were entrusted with authority to make legal and religious decisions. Adin Steinsaltz (1989, 20) portrays the following description of how these schools of higher learning functioned in the second half of the first millennium:[16]

> Usually the students would raise a series of questions before the assembled gathering, questions of interpretation, Halakhik questions, difficulties in various sources, or other problems of logical analysis. These questions would be answered by the head of the yeshivah, but every one of the students had the right to take part in the debate, to raise objections, and answer them, according to his ability. Usually the discussions continued until the problem was clarified or until the assembled scholars decided that they did not have sufficient information to solve it. Questions such as these were sometimes passed from one yeshivah to another, and sometimes from Babylonia to Eretz Israel and back.
>
> (Steinsaltz 1989, 20)

("Halakhik" is derived for the Hebrew word *Halakha*, meaning "to walk" or "to go" and referring to following the Jewish religious laws; "Eretz Israel" is the land of Israel, i.e., Palestine.)

There is no doubt that literacy must have increased among Jews in Babylonia either at the end of antiquity, or in the early Middle Ages, as a yearly event of central importance called *yarchei kallah*, occurring in Babylonia, testifies. Owing to the great extent of the country, opportunities had to be given to those living far from the great *yeshivot*, or busy with agriculture, to take part in the work of the *yeshivot*. *Yarchei kallah* means literally "months of the bride" where the "bride" is probably the *Talmud*. It took place twice a year in the Jewish months of Adar (the month before Passover, at the end of the winter) and Elul (the month before the Jewish new year, at the end of the summer) when the land requires less attention. A precious and interesting account of the formality of this gathering is offered by a Jewish traveler and chronicler, Nathan ben Isaac ha-Kohen ha-Bavli, of the 10th century:

> And this is the order in which they sit. The *rosh yeshivah* stands at the head, and before him are ten men (comprising) what is called the "first row," all facing the *rosh yeshiva*. Of the ten who sit before him, seven of them are *rashe kallot* and three are *ḥaverim*. They are called *rashe kallot* because each one of them is in charge of ten members of the Sanhedrin, and they are the ones called *allufim*.
>
> (Schiffman 1998, 751)

The *rosh yeshiva* is the head of the *yeshiva*, that is, the *gaon*. The seven *rashe kallot* (pl. of rosh *kallah*) in the first row were the highest officers; the *ḥaverim* are associated scholars. The positions of both groups, as Nathan reports, were hereditary, indicating elitism in learning. The seventy *allufim* (masters) imitate with their number the Great Judean Sanhedrin, and this emphasizes once more the link between law school and law courts. The above passage indicates the strict formality of the *yeshiva* at that time, with titles, a hierarchy, and fixed for the prominent scholars. With a bit of imagination, one could find an analogy between the *rosh yeshiva*, the rosh *kallah*, and the *allufim*, respectively, and the rector, dean, and faculties in medieval and modern universities (with the difference that universities normally do not go as far as granting hereditary positions).

Behind all these officers sat, with no fixed place, hundreds of students. Nathan's Hebrew text, published at the end of the 19th century by Adolf Neubauer (1895, 87), librarian at the Bodleian Library, mentions in Hebrew numbers "כד" (24) hundreds. Lawrence Schiffman (1998, 751) has translated the expression as 2,400, but the same expression could also mean "about 4 hundreds." Even 400 is an impressive number.

One of the aims of the meeting was to test these students, and here is an early evidence of examinations:

> When the *rosh yeshivah* wants to test them in their studies, they all meet with him during the four Sabbaths of Adar. He sits and the first row recite before him while the remaining rows listen in silence. When they reach a section requiring comment, they discuss it among themselves while the *rosh yeshivah* listens and considers their words. Then he reads and they are silent, for they know that he has already discerned the matter of their disagreement. When he finishes reading, he expounds the tractate which they studied during the winter, each at home, and in the process he explains what the disciples had disagreed over. Sometimes he asks them the interpretation of laws. They defer to one another and then to the *rosh yeshiva*, asking him the answer. And no one can speak to him until he gives permission. And (then) each one of them speaks

according to his wisdom. And he expatiates on the interpretation of each law until everything is clear to them. When everything is fully clarified, one of the first row arises and publicly lectures on it until everyone understands.

On the fourth Sabbaths the whole Sanhedrin and all the disciples are called, and the *rosh yeshivah* examines each one of them and interrogates them till he discerns who are the more accomplished students. When he notices one whose learning is deficient, he is harsh towards him, diminishes his stipend, rebukes him, indicates those areas in which he is deficient, and warns him that if he repeats his poor performance once more, his stipend will be completely cut off. Therefore, they would sharpen their wits and diligently apply themselves so that they would not make any mistakes before him regarding a matter of law.

(Schiffman 1998, 752)

Learning remained mnemonic, and students studied on grants; their aptitude to receive a grant was determined thorough an examination. These formalities, however, remained within the sphere of devoutness—hardly comparable with the corporatism of medieval universities, let alone the medieval faculties of theology that certainly belonged, of course, to the same sphere. We do not know how early all these were introduced to the *yeshivot*. Some aspects of the *yeshiva* didactics, nevertheless, would be worth considering, even by modern universities. Agassi (1987, 16), confessedly a *yeshiva* drop-out, remarks, "The dedication of the average scholar in a Jewish theological school, even in adolescence, is unbelievable; the most successful, slave-driving, senior high-school classes, even the most taxing colleges, do not achieve anything like it. A 12-hour day of very intense work is a light burden for the theological school students." In modern *yeshivot*, texts are often studied by pairs of students who integrate each other's knowledge; a method that is particularly appropriate as a format for discussion.

CONCLUDING REMARK

Ancient Jewish higher learning concentrated on religious law studies. It consisted in mnemonic learning and, at a higher level, debate.

BIBLIOGRAPHICAL NOTES

The main source of Biblical schooling is, of course, the Bible itself (Old and New Testaments and apocrypha—found, in some editions, in a separate section between the Old and New Testaments) and Josephus Flavius's writing, originally written in Greek. In particular, *War of the Jews, Antiquities of the Jews*, and his *Autobiography*. These texts can be found in English translation online. One should of course consider the historical context in which they were written. There is, then, the enormous rabbinical literature, mostly in Hebrew and Aramaic, which is not easy to read even for native Hebrew speakers and contains relatively little information concerning higher schooling methods and institutions.

Adin Steinsaltz's the *Reference Guide* to the *Talmud* is a helpful tool for the laymen. It offers a clear and concise introduction, including institutions, method of learning, and Talmudic era. It is part of a monumental work that aims at spreading the study of the *Talmud*, whose traditional canonic form, whether in Hebrew or Aramaic, is only readable by experts. Steinsaltz's *Guide* is a laudable contribution from a religious and popular point of view, but uncritical from the historical point of view.

Catherine Hezser's relatively recent *The Social Structure of the Rabbinic Movement in Roman Palestine* and *Jewish Literacy in Roman Palestine* are useful, massive collections of data from primary and secondary sources.

A detailed and interesting study of the *batei midrash* is Gerald Marvin Phillips's doctoral dissertation on rhetoric in the Talmudic schools, submitted back in 1956 and available on the Internet. Phillips's work, however, like many other works on this same topic, looks at the institutions of higher learning from too modern a vantage: half a century ago, historiography was still much positivist.

NOTES

1. Flavius Josephus, The Life of Flavius Josephus, www.sacred-texts.com/jud/josephus/autobiog.htm, accessed February 8, 2012.
2. Ibid.
3. http://www.biblegateway.com/passage/?search=2%20Chronicles%2034&version=NIV, accessed February 2, 2013. The Hebrew terms in brackets added by the author.
4. http://www.biblegateway.com/passage/?search=Nehemiah+8&version=NIV, accessed January 29, 2013.
5. http://www.4thewordofgod.com/maccabes.htm, accessed February 6, 2013.
6. http://lexundria.com/j_bj/2.117-2.166/wst, accessed November 3, 2014.
7. http://www.come-and-hear.com/bababathra/bababathra_21.html#21a_1, accessed June 26, 2013.
8. http://www.chabad.org/library/article_cdo/aid/2099/jewish/Chapter-Five.htm, accessed November 3, 2014.
9. For a collection see Christopher P. Benton, "Some Personal Reflections On Educational Practices Found In Rabbinic Literature," http://www.docbenton.com/reflections.pdf, accessed July 5, 2013.
10. http://www.come-and-hear.com/bababathra/bababathra_21.html#21a_1, accessed June 26, 2013.
11. Ibid.
12. http://www.chabad.org/library/article_cdo/aid/2032/jewish/Chapter-Four.htm, accessed July 5, 2013.
13. "ולא הביישן למד ולא הקפדן מלמד." The exact translation is "and the timid does not learn and the pedant does not teach."

14. For a clear and detailed listing of the various methods of Talmudic hermeneutics for the lay reader, I recommend Adin Steinsaltz's *Reference Guide* to the *Talmud* (1989, 149–154).
15. For further details see the entry "Academies in Babylonia" in the 1906 *Jewish Encyclopedia*. http://www.jewishencyclopedia.com/articles/710-academies-in-babylonia, accessed May 12, 2013.
16. Regrettably, without mentioning a source.

5   The Religious Roots
Medieval Intermezzo

Higher learning in the medieval Islamic world was basically religious

Monotheistic religious education, inaugurated by Judaism, influenced, and was in turn influenced by, other educational traditions, including the classical (polytheistic) tradition. The Early Middle Ages saw the decadence of the Christian West and the flowering of Islamic culture and science. The Jews were scattered in Islamic lands, and Islamic religion and culture was as profoundly influenced by Judaism as it was by Christianity. The heterogeneous Islamic medieval world stretched from Spain to India, and although higher learning in Islamic societies varied in time and from place to place, it contributed considerably to learning—a contribution that can hardly be summarized into a few lines. A few general aspects can, however, be drawn.

In Islam, as in Judaism, study is given a high priority, and there are many similarities between Jewish and Islamic early medieval education. Islam, like Judaism and unlike Catholicism, is a nomocracy: nobody can claim to be above the (religious) law. Both religious educative curricula concentrate upon the law, called by the Muslims *shari'a*—"the way," or "the path"—the same meaning as the Hebrew *halakha*. Much like the Jewish *beth midrash*, the mosque (from *masjid*—place of prostration) served as a place of religious learning and discussion, social contacts, communal organization, or other similar activities.

In its first 150 years, Islam saw many informal personal schools of law or *fiqh* (religious jurisprudence) called *maḏāhib* (plural of *maḏhab*, meaning "doctrine"). The most prominent were in Damascus, Kūfa, and Basra in today's Iraq, and Medina. Nevertheless, by the mid-10th century, only six, each named after a famous jurist of the past, survived and gained recognition in Sunni Islam. Four of them endured to the present.[1] The reason for the disappearance of most of the schools is debated (see Makdisi 1981, 1–9) and exceeds the scope of the current work, which concentrates now on the later higher learning institutions in the Islamic world that contributed so much to the roots of what we call today "science."

Higher learning in Islamic societies, being strictly related to religion, ensued, at least partly, from the patronage of pious rulers, and this resulted primarily in translating and studying classical works, or works from bordering cultures. The relationship between patron and client had already been regulated in Roman society, and in the Islamic world it became a main vehicle for promotion of knowledge, which later contributed considerably to the European Renaissance. Pious purposes and religious learning occasionally engendered profane learning, mainly literary and philological, its development being favored, inter alia, by the absence of a central Islamic religious authority as characteristic of Catholicism, and by the personalistic nature of the teaching.

At least three main Islamic intellectual centers flourished during the early Middle Ages: Baghdad, under the 'Abbāsid caliphate; Córdoba, under the Umayyad's caliphate; and Cairo, under the Fatimid caliphate.

The 'Abbāsids were the descendent of Muhammad's youngest uncle, 'Abbās ibn 'Abd al-Muttalib (566–653), and as such they claimed the title of caliph ("successor to the prophet"). They overthrew the earlier Umayyad caliphate, whose capital was Damascus, and in 762 the 'Abbāsid caliph al-Manṣūr, reigning from 754 to 775, founded the city of Baghdad to become the new capital of the caliphate and a leading intellectual center.

Al-Manṣūr favored astrology and medicine by appointing court astrologers and physicians, and this had consequences for theoretical and technical underpinnings. The dynasty was subject to Persian influence and appealed to non-Arab Muslims (*mawali*) that were marginalized under the previous Umayyad dynasty. While carrying on the Hellenic tradition of collecting books and establishing libraries, it adopted practices from the Sasanian Empire—the last pre-Islamic Persian Empire, surviving till 651—including translating foreign works, primarily Greek. The translation movement gained momentum during the reign of caliph Hārūn ar-Rashīd, from 786 to 809, and culminated under the caliphate of al-Ma'mūn, from 813 to 833. The enterprise was smoothed by the use of paper, which reached the Islamic lands from China during the 8th century. Many foreign works were translated into Arabic from Greek, Sanskrit, Persian, and Syriac. This translation movement of the early 'Abbāsid period, as A.I. Sabra (1930–2013) remarks,

> Was a massive movement which took place in broad daylight under the protection and active patronage of the 'Abbāsid rulers. Indeed, in terms of intensity, scope, concentration and concertedness, it had had no precedent in the history of the Middle East or of the world. Large libraries for books on "the philosophical sciences" (*ḥikma*, or *al-'ulūm al-ḥikmiyya*) were created, embassies were sent out in search of Greek manuscripts, and scholars (Christians and Sabians)[2] were employed to perform the task of translation, all of this at the instigation and with that financial and moral support of the 'Abbāsid caliphs.
>
> (Sabra 1987, 228)

The opportunity given to scholars having different backgrounds to interact openly no doubt broadened their world view and contributed to the development of science. More specifically, as far as learning mathematics in Islamic societies is concerned,

> Elementary texts composed before 1200 reflect the needs of legal schol-ars for some training in arithmetic and their desire to get acquainted with Indian digits and rules for multiplying, dividing, adding, subtract-ing, halving, doubling, and extracting square and cubic roots with them that were introduced in the late eighth or early ninth century from San-skrit sources. Although they are not necessarily presented as classroom texts, they were meant to serve as teaching tools for private studies or in the circles of religious scholars held at mosques. The latter are the only examples of larger public classes known from this period and were focused on religious knowledge. It is unknown how much mathematics was taught in them, and it is unlikely that classes dedicated specifically to one of the mathematical sciences were held at mosques in this period. Mathematical education beyond the basics that were taught during the years of childhood seems to have been delivered by house teachers and itinerant tutors or was acquired as an autodidact.
>
> (Brentjes 2014, 86)

Yet, as Sonja Brentjes (2008, 324) also adds, "Without patronage, the flourishing mathematical cultures of Islamic societies would have been impossible or at least much poorer."

One institution, often mentioned in relation to the 'Abbāsid patronage, although little documented, was Baghdad's *bayt al-ḥikma* (house of wis-dom). It, too, had the function of a mosque (BenAicha 1986, 256) and is sometimes presented as a center of translation, although it may have been only a collection of rare books (Balty-Guesdon 1992); Dimitri Gutas believes that the term is the translation of the Sasanian designation for library, and in Sasanian times the *buyūt al-ḥikma* (pl. of *bayt al-ḥikma*) functioned as ide-alized national archives (Gutas 1998, 54 and 56). Under Hārūn ar-Rashīd, the *bayt al-ḥikma* was probably used only by the caliph, the people around him, and a few translators.

Al-Ma'mūn may have opened the *bayt al-ḥikma* to a larger public of scholars; he also employed—in the library or outside it—the outstanding Per-sian mathematician (more precisely, astrologer) Abū 'Abdallāh Muḥammad ibn Mūsā al-Khwārizmī (c. 780–850; Gutas 1998, 58). Al-Khwārizmī's *'Al-Kitāb al-mukhtaṣar fī ḥisāb al-jabr wa-l-muqābala'* (The Compendious Book on Calculation by Completion and Balancing), written with al-Ma'mūn's encouragement as a standard work on calculation, originated the term *algebra* (from *al-jabr*, which means "completion," or "to restore," indicating carrying a term from one portion of an equation to another with the relative change of sign); al-Khwārizmī's name is also at the origin of the mod-ern mathematical term *algorithm*. There was, then, the Neoplatonic philoso-pher al-Kindī (c. 801–873), who was "not only eager and able to absorb and expound an astounding amount of Greek science and philosophy, but also wrote openly and forcefully against religious bigotry and intellectual insularity" (Sabra 1987, 226). One famous 9th-century Baghdadian translator was Thābit ibn Qurra (826–901), who was a court astrologer and proficient in geometry, philosophy, and even medicine; he translated or edited from Greek into Arabic works of Archimedes, Euclid, Ptolemy, and Apollonius (c. 262 BC–c. 190 BC—the Greek geometer and astronomer known for his works on conic sections). One could mention many other intellectuals: Al-Ma'mūn also set up an observatory, Shamsiya, near Baghdad. Under caliph al-Muta'sim billāh (r. 833–834), at the last, the teaching of mathematical sciences was introduced at courts, and the caliph's grandson, Ahmad, to become caliph al-Musta'īn billāh (r. 862–866), was sent to study under al-Kindiī (Brentjes 2014, 87).

One other institution of higher learning under the 'Abbāsids (at least, so it seems)—not explicitly religious, but supported by pious endowments—were the hospitals, applying Hellenic, Persian, and Indian Medicine (Bennison 2011, 90). Under the aegis of al-Manṣūr, a community of practicing physicians, many of them of Nestorian origin, is said to have emerged in Baghdad. The Nestorians were Christians, originally follower of Nestorius, Patriarch of Constantinople between 428 and 431, who emphasized the disunion between the human and the divine natures of Jesus; their view was condemned by the First Council of Ephesus (431), and many Nestorians relocated to Sasanid Persia and were later instrumental in spreading Hellenic philosophy and medicine.

Although the origin of hospitals in the Islamic world, in general, is not quite clear, they were widespread in the 'Abbāsids caliphate: Harun al-Rashid founded a hospital that served also a center for translations (Balty-Guesdon 1992, 146). During the caliphate of al- Ma'mūn, vizier Yaḥyá ibn Khalid (d. 806; "vizier" is a high-ranking political advisor or minister), belonging to the influential family of the Barmakids, is reported to have paid physicians from India to run a hospital, translate medical books from Sanskrit into Arabic, and bring Indian pharmaceutical plants and drugs to Baghdad (Brentjes & Morrison 2010, 569). Originally from Buddhist or Zoroastrian clergy, they served and advised the 'Abbāsid dynasty for generations (Zoroaster was the predominant Persian religion before Islam.). The Barmakids arranged *majālis* (places of sitting, salons) where courtiers and scholars, including Jews, Christians, and Zoroastrians, met to discuss religion and philosophy. Hospitals, too, had their *majālis*, at times acting as rudimentary medical schools, and one way to study medicine was to attend classes that physicians held there or in their own homes (Huff 2003, 161–162; Leiser 1983). The texts commonly used were Galen's works, and the method of studies was the traditional with all of its shortcomings: Gary Leiser (1983, 51) comments, "Many teachers made no effort to explain the medical texts to their students. Instead, they just listened as the students recited the texts before them. As a result, the students had no understanding of what they

had read." Occasionally students were examined, as a warranty of quality, although there is no evidence of a title being granted analogous to the
modern MD (from the Latin *Medicinae Doctor*, teacher of medicine, today
the university degree for physicians). The most common avenue to become
a physician remained to belong to a physician's family. Physicians could
also be self-taught through study of the many available medical texts. Memorization always remained the primary method of study. One prominent
partly self-taught physician was ibn Sīnā (Avicenna, c. 980–1037), the Persian polymath whose encyclopedic *The Book of Healing* became a standard
medical text in European medieval universities.

What was the drive behind the court-sponsored patronage and translation enterprise? Traditionally, the drive was both utilitarian and religious:

> The motives and objectives that caused this massive cross-cultural
> transfer of knowledge under the 'Abbāsid caliphs and their courtiers are
> seen as answering the practical needs of the new dynasty, among them
> astrological and medical concerns. Ritual duties of the Muslim com-
> munity such as praying at particular times and in specific directions are
> thought to have inspired an interest in various mathematical disciplines.
> Religious debates that brought together members of various Christian
> Churches, Manichaean dualists, Mazdakites, Jews and adherents to
> various Muslim factions left Muslim disputants in an uncomfortable
> position, as they were unfamiliar with the various tools of pre-Islamic
> philosophical and theological debates. It has been argued that a handful
> of 'Abbāsid caliphs promoted enlightened, tolerant and rational values
> in a politics that was opposed to obscurantism and liberalism.
>
> (Brentjes & Morrison 2010, 565)

Yet Gutas (1998) and Brentjes (Brentjes & Morrison 2010) also suggest
consideration of a broader political context related to the end of Byzan-
tium's oppressive religious control on one hand, and the endeavor to restore
pre-Islamic Persian splendor favored by the 'Abbāsids on the other.

The *majālis*, bringing together people of differing views to attend debates
in quasi "academic freedom," were not only an important institution of
contemporary higher learning in medieval Islamic countries; they were, inter
alia, a vehicle for scholars to approach patrons to find board and lodging
and, eventually, employment as physicians, royal scholars, ministers, and
other positions (Benninson 2011, 181–182).

Finally, Islamic higher learning, much like its Jewish contemporary,
granted certificates, which may be considered a forerunner of the later university degree. "Sooner or later the question was bound to arise whether
a scholar's claim to have heard certain teachers was genuine . . . It may be
assumed that the first solution of this problem was a written license stat-
ing that a man had studied certain books under a given teacher. But the
license (*ijāza*) soon obtained an independent existence" (Tritton 1957, 40).

There were different types of *ijāza*, looking more like contracts between teacher and students, as assurance of validity and quality of the material taught, rather than official certificates; Arthur Tritton produces many examples (Tritton 1957, 44–45; for an early one from year 976, see p. 40). Yet, as Daphna Ephrat (2000, 69) is convinced, "It never developed into an institutionalized degree, such as the *licentia docendi* granted by European universities."

As early as in the 9th century, the 'Abbāsid caliphate began to crumble: centrifugal tendencies asserted themselves, particularly in Persia. In the 10th century the caliphate came under the custody of the Daylamite Buyids. The Daylamites were Iranian People inhabiting the mountainous regions on the southern shore of the Caspian Sea, employed as soldiers from the time of the Sasanian Empire. They long resisted the Arab conquest and Islamization, and in the 930s, the Daylamite Buyid—an Islamic Shi'ite dynasty—managed to gain control over much of modern-day Iran, until the coming of the Seljuks—nomadic Turkish warriors from central Asia—in the 11th century. The caliphate came under the custody of the latter, although it acquired a stronger and more independent status. The 'Abbāsid caliphate lasted till the Mongol's conquest of Baghdad in 1258.

Besides Baghdad, Córdoba in Spain (more precisely, al-Andalus, or Islamic Iberia) was an important intellectual center. It was subject to the Umayyads who, after having been overthrown from Damascus, reestablished their rule in this part of the world. The 10th-century library of al-Ḥakam II (r. 961–976), second caliph of Córdoba, contained extensive translations from Greek into Arabic.

In North Africa, too, seeds of Islamic higher educational institutions began springing up. During the 8th and 9th centuries, the mosque of Qayrawan, in today's Tunisia, grew to become a center of learning (BenAicha 1986, 256–257). In 859 a wealthy heiress, probably originating from Qayrawan, founded a mosque in Fez, Morocco, and an elite of Qarawani scholars moved to teach there. The Fez mosque/school—originally a satellite mosque of Qayrawan—entered the Guinness Book of Records, rightly or wrongly, as the oldest "university" in the world, even if only in 1963 was it officially declared the state university of "al-Qarawiyyin."

In 969, the city of Cairo was founded, north of Fustāt, as capital of the Fatimid caliphate, ruling North Africa. The Fatimid claimed to be descended of Fatimah, daughter of Muhammad, and of Ali, son-in-law and cousin of Muhammad, and belonged to the Ismaili (transliterated Ismā'īlī) Shi'a school of faith ("Shī'ah" meaning "followers," "faction," or "party" of Ali; Ismailism is a branch of the Shī'ah).

> Knowledge (*'ilm*) and wisdom (*ḥikma*) are, according to Ismaili belief, gifts from God, revealed to humanity through His prophets. God has successively dispatched six prophets bearing a law (*sharī'a*): Adam, Noah, Abraham, Moses, Jesus Christ and Muḥammad. These prophets

are called 'speakers' (*nāṭiq*), because they talk to men, proclaiming to them a *sharī'a*, an exoteric (*ẓāhir*) law with its commandments and prohibitions, its ritual obligations and legal definitions. By the side of each of these speaker-prophets stands an authorized representative (*waṣī* or *asās*) who knows and teaches the eternally immutable 'esoteric meaning' (*bāṭin*) of all these prescriptions and regulations—though only to a small number of the elect.

(Halm 1997, 17)

Just as Moses and Jesus had, respectively, Aaron and Peter as "authorized representative," Ali, according to the Ismailis, was Muhammad's *waṣī* or *asās*. Ali's descendants are the true imams (leaders) of the Islamic community: they alone know and transmit the esoteric meaning of the divine revelation. The Ismailis may have been one of the sources to the later European Renaissance esoteric practices (Eamon 1991, 335).

In 972 the Fatimids inaugurated the al-Zahra ("the Luminous") mosque in Cairo in honor of Fatimah. They, too, encouraged scholars and jurists to have their study circles and gatherings in or around the mosque, turning it into a study center, and eventually it became a university in modern times. The Fatimids, too, collected books on a variety of subjects, and their libraries attracted the attention of scholars from across the world. They encouraged both exoteric and esoteric knowledge and wisdom, and were, initially at least, tolerant towards non-Ismaili Muslims as well as towards Jews and Christians. In this context, Goitein reports interesting relations between Jewish and Islamic higher learning, testified by the Cairo Geniza:

In the last third of the tenth century, when an eclipse in the fortunes of the Palestinian yeshiva coincided with the prominence in Egypt of a great scholar of Iraqian training, Shemarya b. Elhanan (whose father had already been active there), a midrāsh, or college, was formed in Fustat by him. This development may have been connected with the advent of the Fatimids and the fact that at that time Egypt had become the center of a great empire, as well as the fountainhead of intensive, thoroughly organized religious propaganda on behalf of the Isma'ilīs, the ruling sect. We should keep in mind that, at the same time and place, a former Jew from Iraq, Ya'qūb (Jacob) Ibn Killis, converted to Islam and Fatimid vizier, personally presided over the establishment of the nucleus of what was later to become the famed Muslim University of al-Azhar.

(Goitein 1971, 202)

Cairo, too, had its House of Wisdom (*dar al-ḥikma* or *dar al-'ilm*) built around 1005 as a library and converted by the much discussed caliph al-Ḥākim (r. 996–1021; he gave the order, among others, for the destruction of the Church of the Holy Sepulchre in Jerusalem) to a center of studies.

The *dar al-ḥikma* aimed primarily at training Ismaili teacher-propagandists and was more religiously focused than the Baghdad *bayt al-ḥikma*, although it left some room to "foreign" philosophical sciences (Benninson 2011, 89).

In the second half of the 12th century the Fatimid caliphate declined rapidly, and in 1171 Egypt was taken over by Saladin (Ṣalāḥ ad-Dīn (c. 1138–1193), of Kurdish origins), who was proclaimed the "Sultan of Egypt and Syria" by the 'Abbāsid caliph al-Mustadī' and founded the Ayyubid dynasty (the Ayyubids were Saladin's family), which ruled much of the Middle East during the 12th and 13th centuries.

The most interesting higher learning Islamic formative institutions were, nonetheless, the *madrasa*s (sing. of *madrasa*—school—a term with the Semitic roots d-r-s, as *beth midrash*, meaning the place to study) from the 10th century onwards. The origin of the madrasa is debated (Leiser 1986, 16–17). It has been suggested, among other possibilities, that *madrasa*s were inspired by Buddhist monasteries; that they were stimulated by the *dar al-'ilm*; and that they were a logical outgrowth of the mosque as a rudimental teaching place—a teaching that became more permanent, although always informal. *Madrasa*s were above all religious institutions of higher learning.

Madrasas, like Yeshivot and some early European universities, were (religious) law schools

The religious roots of Islamic higher education are, as the Jewish ones, in the centrality of (religious) law and its jurisprudence. And the law was to become one of the first teachings in European universities. One other phenomena that characterized both Islamic and European medieval higher education was the wandering of scholars—including scribes, physicians, poets, and students—from place to place, looking for better opportunities (Ephrat 2000, 33–34). This was also a support to the diffusion of knowledge in an era in which speech and manuscripts were the only instruments available.

As jurisprudence developed, more expertise was required, and, much as in medieval Europe around cathedrals, communities of students and teachers sprang up in proximity to mosques where legal studies took place, often supported by pious endowments (*waqf*, meaning "confinement and prohibition," i.e., detention by the beneficiary). The local *khans* (inns) played an important logistical role, according to George Makdisi (1920–2002). In Makdisi's words, in the *madrasa*

> was a natural development of two previous institutions: the masjid, in its role as a college of law, and its nearby khan, as the residence of the law students in attendance. The development of this college was made in three stages: from the masjid, to the masjid-khan complex, to the *madrasa*. The masjid involved in this development was that in which the

teaching was devoted to law as its primary subject. The basic law course of the masjid, usually lasting a period of four years, required a place of residence for the law students who came to it from out-of-town; whence the development of the masjid-khan complex. From this complex to the *madrasa* there was but one simple step. The essential difference between the second and third stage of development is to be found in the legal status of the masjid and that of the *madrasa*. Both institutions were based on the law of waqf, charitable trust [. . .] the masjid, once instituted as waqf, became free of its founder's control. Its waqf was said to be a *waqf taḥrīr*, 'a waqf of emancipation'. The relationship between it and its founder was thus likened to that existing between an emancipated slave and the emancipating master who relinquishes his rights over him. In contrast, the *madrasa* came under the control of its founder, and that of his descendants, in perpetuity, if he so desired.

(Makdisi, 1981, 27–28)

Indeed, during the 10th and 11th century, *waqf* donations were increasingly assigned for foundations. The conditions of the *waqf* foundations varied: sometimes they provided a salary for teachers and stipends for students, sometimes only salaries for teachers, and sometimes only the physical facilities themselves (Sizgorich n.d.). The main causes for patronage became the need for political and spiritual advice and self-expression, and perhaps even the need to prepare for the afterlife as well as for finding allies among the civilian elites, economic necessities like water works, physical necessities like medicine, and vows to God or a saint.

In the middle of the 11th century *madrasa*s were founded in Iran and Iraq by Seljuk princes, emirs, or their administrative officers like viziers. In particular, Abū ʿAlī al-Ḥasan al-Ṭūsī, Niẓām al-Mulk (1018–1092), a Seljuk scholar and all powerful vizier, founded a number of *madrasa*s that took his name: *Niẓāmiyya*; and for Sabra (1987, 233), "He was directly responsible for initiating the system of *madrasa*s." Among these *madrasa*s, the most famous was the Baghdad *al-Niẓāmiyya*, founded in 1065 or 1066. It

truly signaled a shift from the other institutions (mosques, palaces, bookstores, and scholars' houses) to organized, government-supported schools. From that time onward, public schools spread throughout the Muslim world in an organized manner, with government authorities, merchants, and upper class individuals competing to establish schools. Thus, a new and prosperous era in the development of Islamic schools had been ushered in.

(al-Otaibi & Rashid 1997, 8)

The philosopher and theologian al-Ghazālī (1058–1111), author of *The Incoherence of the Philosophers*, which denounces Greek philosophy and its Muslim followers, taught there for a few years at the end of the century.

In Egypt, due to the lingering Fatimid regime,

> The *madrasa* was a relative latecomer. Egypt's first *Madrasa* may have been established as early as 1097, and several founded by Sunni ministers to the late Fatimid caliphs were already functioning in Alexandria by the time Saladin abolished the Shi'i caliphate and established his Ayyubid dynasty. Saladin did establish the first *madrasa* in the Egyptian capital itself, or, more precisely, in the neighboring city of al-Fusṭāṭ, in 1170. Under his patronage, and that of his successors and other individuals, at least thirty-two *madrasa*s for instruction in Islamic jurisprudence and related subjects were founded in the urban metropolis of Cairo/al-Fusṭāṭ by the time of the Mamluk coup against the last Egyptian Ayyubid.
>
> Thus, by the middle of the thirteenth century the *madrasa* had established itself as the primary forum for religious education throughout the Islamic Middle East, and in particular in the Egyptian and Syrian provinces of the Mamluk empire. The institution, like higher education itself, was a largely urban phenomenon. The principal cities of Syria—Damascus, Aleppo, Jerusalem, and others—were home to numerous *madrasa*s and to many of the most reputable scholars of the later Middle Ages, and consequently attracted students from other Muslim countries.
>
> (Berkey 1992, 8–9)

The main subject of study in the *madrasa*s was the *fiqh*, and as such they can be considered as law schools or, more precisely, colleges specialized in the teaching of law. The alien Turkish regime needed bureaucrats skilled in the Arabic language and Islamic law, and *madrasa*s could provide these experts. As Ira Lapidus (2002, 192) remarks, the Seljuks had no experience in ruling an agriculturally based empire, and here scribes and law experts schooled in *madrasa*s came to be vital.

Madrasa teaching was personalistic and informal: students selected their *madrasa*s according to the teachers, the stipend they could get, or the reputation of the city in general. There, a master would lecture, often in a form of dictation, on topics such as jurisprudence, religious doctrine, and traditions of the Prophet. He could recite from memory his own works or works of other masters and might have been accompanied by an assistant, who wrote down his words. He is described as sitting on a cushion cross-legged or a chair leaning against a wall or pillar, surrounded by students sitting cross-legged in a semicircle, taking notes; the most qualified students were seated in proximity of the teacher (Ephrat 2000, 77). Beside dictation, studying could be mnemonic, through reading, recitation, or even discussion. Normally each study center had its canonical texts, and these could differ from place to place. Submission and deference were typical, although some teachers encouraged their students to ask questions, criticize, and point out errors.

The student, from beginning to end, had to continue to develop a strong memory, learn how to stock it carefully with the necessary stores, and so arrange and classify them there as to be able to retrieve them with the least possible hesitation, drawing upon the memory's treasures at will. For advocacy was a completely oral exercise. There was no time for reference to sources, no time for that deliberation one has when writing, no opportunity to draft and redraft before delivering the final product. All deliberation had to be done beforehand, and the material mastered definitively for instant recall, in preparation for the supreme encounter with the adversary, at which time there would be no margin left for error, no quarter asked and none given. Reputations were made and unmade; careers hung in the balance.

In medieval Islam, the achievement of consensus (ijma')[3] was made possible by the absence of disagreement (khilaf). Ijma' was thus arrived at by a system of elimination. Eliminate khilaf and you have ijma'. Those who sought the achievement of consensus had therefore to see to the elimination of disagreement. This was the goal to be achieved; it was to be achieved in one of two ways: by winning the adversary over to one's side, or by reducing him to silence.

The advocate's training thus revolved around khilaf. The object of training was to learn how to meet all possible objections to one's thesis. His training was dominated by two major initial concerns: (1) to commit to memory an ever-growing repertoire of questions still being disputed, and (2) to learn and practice the art of disputation, or argumentation, with special emphasis on how to ask questions and how to answer them.

(Makdisi 1981, 112)

In some *madrasa*s students got an *ijāza*, which entitled them to teach what they learned.

Was the Islamic madrasa *influenced by the Jewish (Gaonic)* yeshiva?

Although the Gaonic *yeshivot* had a small sphere of influence in a huge Islamic world and flourished before the *madrasa*s, there are, indeed, a number of general similarities: both were higher learning religious law schools; both bridged between an oral and a literate culture; in both the teacher's authority had a central importance, and occasionally students sat according to their rank; and both could grant a certificate of proficiency. There were, nevertheless, also significant differences, as discussed by Daphna Ephrat and Yaakov Elman (2000). The *yeshivot* (much like the European universities, with their faculties, deans, rectors, chancellors, or presidents) had a much more rigid, hierarchical, elitist, social structure; here the sitting pattern in *yeshivot*, such as Sura and Pumbedita, is exemplary. Moreover, a student of a *yeshiva* could hardly establish his own *yeshiva*, whereas a *madrasa* student holding an *ijāza* could found his own school. The *yeshivot*, unlike the *madrasa*s, could act as courts. Additionally, whereas in Jewish (and Greek) higher level schools one studied for the sake of studying, "the thrust of the educational system in the college of law," as Makdisi (1981, 111–112) remarks, "The *madrasa* or its precursor, the masjid-khan complex, was directed toward the training of the advocate," that is, according to Makdisi, it was more vocational. The *yeshivot* were also more orally oriented than the *madrasa*s. Despite that the *yeshivot* most probably did not influence directly the *madrasa*s, the general patterns, that is, religion, law studies, and license granting, are remarkable as general features of medieval higher learning. There have undoubtedly always been contacts and mutual influences between religions, and even more interesting are the similarities between the *yeshivot* and the medieval universities, which will be discussed later.

According to Makdisi, *madrasa*s had characteristics similar to those of the early European university colleges (as opposed to universities): "the *madrasa* was the embodiment of Islam's ideal religious science, law, and of Islam's ideal religious orientation, traditionalism; and that law and traditionalism combined to produce the scholastic method which was the peculiar product of the Middle Ages" (Makdisi 1981, xiii).

To be sure, "scholastic method" was practiced in medieval European higher learning and was based on the authority of the Latin Fathers and of Aristotle and his commentators. Yet, if one considers it, as Makdisi does, more generally as a detailed and critical examination of texts, one could say that it was practiced both in *yeshivot* and *madrasa*s (critical examinations of texts were typical for Neoplatonic schools as well as for oriental Christian teaching and include writing commentaries of different kinds). "Traditionalism," however, means here something more specific: the preference of the *'Ulum an-naqliyya*—that is, the transmission, preferably orally, of religious Islamic knowledge, mostly the saying of Muhammad or other religious authorities, such as the *hadith*. At some point, when the various types of knowledge were classified and re-classified, these were defined to embrace also exegesis and law. Their opposites, however, were the rational sciences, *'Ulum al-'aqliyya*, and the mathematical and experimental sciences. The former contained religious and non-religious disciplines, in particular the (theoretical) fundaments of *din* (faith/religion) and law, logic, various parts of philosophy (depending on region and period), and philological disciplines. The latter could contain various other branches, such as magic squares, architecture, and later other fields of the more occult.

In summation: although there may be no direct evidence to this claim, there is little doubt of a thread running from the Jewish *beth midrash* and *yeshiva*, through the Islamic *madrasa* to the European medieval university. In ancient times and in the early Middle Ages (unlike in modern higher education) Greeks, Jews, and eventually Muslims likewise studied in a religious context mainly for the sake of studying and of spiritual development. In Islamic lands, however, one can find seeds of vocational study in law as students studied to become jurisconsults.

> But Muslim education was not all there was to education in Islam. Institutionalized learning was not all the learning available. Philosophy, philosophical or rationalist kalam-theology, mathematics, medicine, and the natural sciences, that is those sciences referred to as the ancient, or foreign sciences, as well as all fields not falling under the category of the Islamic sciences and their ancillaries, were sought outside of these institutions, in the homes of scholars, in the hospitals, in the regular institutions, under the cover of other fields such as hadith or medicine.
>
> (Makdisi 1981, 281–282)

These more open institutions, mainly libraries and hospitals, contributed the seeds of the Renaissance and the renaissance of science. There is, of course, much more to say about higher education in Islamic countries after the 13th century. Let me, however, turn to learning in the Christian world before the appearance of the first universities.

In the Christian world, in a society that was almost completely illiterate, the church was the only institution able to conserve and pass on knowledge

Whereas classic higher knowledge was translated, preserved, taught, and improved in the Islamic world, it was nearly totally lost in that part of Europe that was once the Western Roman Empire. Only fragments were preserved, much thanks to a few philosophers and theologians, the most prominent being Severinus Boëthius (in short Boethius) in the early 6th century.

Boethius was born in Rome to a noble Christian family during the last years of the Western Roman Empire. He became a Roman senator, consul (the honorary position heading the Senate), and, around 520, *magister officiorum* (head of the court) to Theoderic the Great, king of the Ostrogoths—the rulers of Italy from 493 to 553, with Ravenna on the Adriatic (the last capital of the Roman Empire) as capital. Eventually Boethius lost the favor of Theodoric, perhaps out of a suspicion that he was in sympathy with Justinian, emperor of the Eastern Empire. Boethius was imprisoned, most probably in Pavia, and executed around 525.

In prison, Boethius composed his most famous work, the *Consolation of Philosophy*, a treatise on issues such as fortune and death, which became popular and influential. Boethius also wrote on theology and contributed to the preservation of Hellenic philosophy and science in Western Europe by writing on the *quadrivium* and on logic. Part of his works was lost, but his works in arithmetic, music, and logic survived and became a reference point for medieval higher learning.

Boethius was not alone, although the number of intellectuals who produced higher learning was scant. Boethius's successor as *magister officiorum* to Theoderic, Cassiodorus (whose full name was Flavius Magnus Aurelius Cassiodorus Senator, c. 485–585), too, contributed to the study of the liberal arts as a tool to better understanding the Bible. After spending a period in Constantinople, he founded, on his family estate in the south of Italy, the monastery of *Vivarium*, so named for the local fishponds, and initiated the tradition of rigorous copying of manuscripts.

One other contemporary theologian and philosopher who contributed to the preservation of knowledge was Isidore of Seville (c. 560–636), church Father, considered the last scholar of the ancient world and the father of European culture. As archbishop of Seville he was involved in the conversion of the royal Visigoth Arians to Catholicism. (The Arians were the followers of Arius (c. 250–336), a Christian presbyter in Alexandria who denied the divinity of Jesus Christ.) Isidore compiled, under the title *Etymologiae*, later also called *Origines*, a summa of all knowledge that became popular in the Middle Ages and the Renaissance. It included, among other subjects, the liberal arts and medicine, although Latinist Andrew West (1853–1943) comments that "his huge book is of course utterly without original value and so full of absurdities and puerilities that it may be considered as an index of the retrogression in learning that had set in" (West 1971, 26). The science may have been poor, but religion remained omnipresent. In general, the early medieval Western Christian world was almost completely illiterate.

From the institutional point of view, the Catholic church was the only institution able to conserve and transmit knowledge, and the little education available was primarily of a religious nature. To encourage devoutness and also maintain institutional stability, the Council of Vaison in 529 decreed that parish priests could host unmarried young men and teach them the Psalms and divine law so as to procure for themselves worthy successors. Schools were founded adjacent to cathedrals in order to prepare the future clergy. Study was often only elementary and limited principally to reading, writing, choral music, and arithmetic—subjects that, nevertheless, at that time were attributed much greater dignity than they are today.

In 826, Pope Eugene II (r. 824–827) held a council at Rome, ordering bishops to appoint in all diocese and parishes masters able to teach the liberal arts and sacred dogma (two types of teachings that may contradict each other).

A second educational institution during the early Middle Ages was monasticism; at that time it went well beyond the religious sphere and became an economic, cultural, and social bulwark in the midst of the general insecurity that reigned at the time. Best known are the Benedictine abbeys at Monte-cassino (founded around 520) and Bobbio (614) in Italy, Fulda in Germany (714), and St. Gallen in today's Switzerland (719). The monasteries conserved part of the cultural patrimony of antiquity, above all thanks to their activities of copying manuscripts in the wake of Cassiodorus's *Vivarium*. Schools were essential for this activity, as they were for preparing new generations of monks. The monastic schools did more than just prepare new novices; on occasion, lay people were admitted and non-religious subjects were taught.

Under Charlemagne, schools combining classical and religious education were founded

The first signs of cultural rebirth and a renewal of education in the Christian world appeared under Charlemagne, who became emperor in 800. Charlemagne aspired to create a center to replace both Rome and Athens. More education was necessary, and the Greek model of instruction returned to the fore. Charlemagne gathered the best scholars of the time at his court, forming the *Schola palatina*. The most prominent among them was the Benedictine Alcuin of York (c. 732–804), an English teacher and theologian, head of the York Cathedral school, with its rich library (Nonn 2012, 21–24). The idea was, above all, to give a fitting education to the emperor, his family, and his courtiers.

Charlemagne also encouraged the foundation of elementary schools in convents and cathedrals. The Palace School was the seed of a movement that fostered the rebirth of culture and education in the West.

> The secular state helped the Church to impose upon herself a reform beneficial to both; and by virtue of their joint efforts Europe's first large-scale educational system came into being. It was markedly Christian in tone and could hardly be otherwise since it was organized round the monasteries and cathedrals. Charlemagne, however, was sufficiently important to demand attention for his particular needs, and the curriculum which came to be introduced in his empire was in consequence somewhat wider in scope than the one the Church had built entirely to its pattern in England.
>
> (Bolgar 1958, 110)

The *curriculum* included the teaching of the seven liberal arts, which Alcuin considered "the seven pillars of wisdom." It is sometimes considered, more wrongly than rightly, the seed of a university.

Charlemagne's cultural and educational enterprise was imitated by Alfred the Great (r. 871–899) of Wessex, the king who defeated the Vikings.

His biographer, Asser (d. 908/909), a Welsh monk who became Bishop of Sherborne, reports the following:

> With wonderful affection he cherished his bishops and the entire clergy, his ealdormen and nobles, his officials as well as all his associates. Nor, in the midst of other affairs, did he cease from personally giving, by day and night, instruction in all virtuous behaviour and tutelage in literacy to their sons, who were being brought up in the royal household [. . .]. He used to moan and sigh continually because Almighty God had created him lacking in divine learning and knowledge of the liberal arts [. . .]. At that point God [. . .] sent some comforts for this royal intentions—certain luminaries, as it were: Wertfeth, the bishop of Worcester [. . .], then Plegmund, archbishop of Canterbury [. . .]; and also Æthelstan and Werwulf, both priests and chaplains [. . .]. However, since the royal 'greed' (which was entirely praiseworthy!) in this respect was not yet satisfied, he sent messengers across the sea to Gaul to seek instructors.
>
> (Keynes & Lapidge 1985, 91–93)

All the educators were, of course, clerics.

The expansion of dialectic, however, set the basis for future university curricula: Greek liberal arts combined with theology (the *quadrivium*, however, comprising four disciplines based on numbers, also included the occult arts; thus, it was not always viewed with favor in Christian medieval culture). The model of the Palace School spread: it was applied at religious schools, such as those in Tours, Laon, Orléans, and Fulda, creating a load-bearing structure for higher learning. The framework was always religious, even though the cultural center of gravity was gradually shifting from the monasteries to the cathedral schools. This led, between the 10th and the 11th centuries, to the birth of Scholastics. The term derives from *scholasticus,* the head of the cathedral school who became a teacher of the liberal arts who would read out loud to his students from texts that were generally taken from Aristotle's *Logica* and then comment there upon. The *scholasticus* could grant a *licentia docendi* (license to teach), in the territory of the dioceses he belonged to, to candidates that had passed a test, taken oath of allegiance, and paid a fee. The *licentia docendi* may be considered the earliest form of European degree. It is, interestingly enough, much of the same kind granted by Jewish and Islamic higher learning institutions.

Later, theologians like adventurous Peter Abelard (1079–1142) or Hughes (1096–1141), *scholasticus* of the Abbey of St. Victor in Paris, opened new horizons by applying the liberal arts to rationalizing theology in what one could call an interdisciplinary approach. Higher learning was taught, above all, in cathedral schools; for instance, a prominent cathedral school is said to have operated in Chartres, hosting a series of scholars studying in philosophy and the liberal arts (Jeauneau 2009). Demand for knowledge grew in the 12th century, and new schools were founded, especially in cities. New-old knowledge trickled to Europe from the 9th century onward, first slowly, only to become substantial during the latter part of the 10th century. The vehicle consisted of diversity of people: among them Muslim travelers (also following the Fatimid conquest of Sicily, a process that began in 827 and lasted until 902), merchants, Christian groups, Jewish individuals and networks, monks, and so forth, first in translations into Hebrew and Latin and then in vernacular. The stage was ready for the birth of universities.

CONCLUDING REMARK

Religion has dominated higher education ever since ancient times, and the aim of European higher education, shortly before the birth of the university, was oriented toward nurturing devoutness. The seeds of more earthly learning nevertheless started emerging or reemerging. Beyond devoutness, (religious) law was learned as a profession, and other subjects such as the liberal arts and medicine were not necessarily in harmony with religion, resulting in the root of the tension and confusion that still surround universities today.

BIBLIOGRAPHICAL NOTES

Al-Khalili's recent *Pathfinders: The Golden Age of Arabic Science* (2010) is an overall popular presentation of the history of Islamic science. Goitein's *A Mediterranean Society* (1971) is a broad study in six volumes of the medieval "Mediterranean Society" under Islamic rule as reflected in the Cairo Geniza; vol. 2 deals with the social and institutional setting. Makdisi's *The Rise of Colleges: Institutions of Learning in Islam and the West* (1981) offers a detailed presentation of the Islamic *madrasa*, and he, too, emphasizes the similarity between these Islamic schools and European universities. The mass of details does not make the book an easy read, in particular for those who are unfamiliar with the context. An excellent appendix at the end of the book reviews previous scholarship. Totah's *The Contribution of the Arabs to Education* is older (1926) and shorter, but reader friendly. Several recent articles by Sonja Brentjes (see Bibliography) convey an up-to-date clear, detailed, and well-documented picture of the relation between Islamic higher learning in the Middle Ages and its institutional frame. Brentjes highlights the importance of patronage and rejects Makdisi's opinion that rational sciences were marginal in the *madrasa*'s curriculum.

Bolgar's *The Classical Heritage and its Beneficiaries* (1958) is a detailed history of the intellectual developments in the Middle Ages and Renaissance in the West, concentrating on individuals rather than institutions. Alcuin and his scholastic reform are presented clearly and concisely in West, *Alcuin and the Rise of Christian Schools*, originally published in 1892.

NOTES

1. See Makdisi 1981, p. 2, and http://science.jrank.org/pages/9936/Law-Islamic-Legal-Literature-Institutions.html, accessed April 16, 2013.
2. It is difficult to say in a few words who the Sabians were. They are mentioned in the Quran as a non-Muslim sect having a faith revealed by the true God, like the Jews and the Christians, but they were not monotheistic, having temples of the moon and Venus, at the very least.
3. The precise diacritic spelling would be *ijmā'*.

6   The Birth of the University

The university arose out of utilitarian needs

Only in this chapter is the university itself finally treated. Let the substance of the five previous chapters serve as an indication of the extent to which the university remains permeated with previous cults and cultures.

Various classical and religious higher learning institutions profoundly influenced the university up until modern times. Yet many of these early institutions did not teach professions. On the contrary, in the classical world, the fortunate few who had access to higher learning normally did not need to work. In the ancient Jewish world—and, to a great extent, in today's *yeshivot*—the purpose of learning was, and remains, spiritual growth. To be sure, one finds seeds of vocational higher learning in medieval Islamic higher education: students delved into the *fiqh* to become jurisconsults, yet the aim of Islamic medieval higher learning, too, remained basically spiritual. The study of medicine is, of course, an exception.

The university arose in European medieval cities with a more utilitarian goal, stressing "having" rather than "being." Fortunately the latter, nobler, goal remained present, although its blending with more "earthly" purposes became a source of confusion. Whereas vocational teachings, such as law or medicine, may provide a living, the liberal arts, aimed at developing virtues, may disappoint those expecting lucrative results. To be sure, broadening the horizons of future lawyers, physicians, or priests could, without doubt, help. Subjects such as grammar and rhetoric are particularly useful for the legal profession, as they are for holding sermons, or for politicians. Nonetheless, the medieval liberal arts remain, to quote Aristotle, "liberal or noble" rather than "useful or necessary"—in contradiction to the university's initial utilitarian goals. Putting the liberal arts in a guild-like framework may open the way to a deviation from ideals ending in corruption. The story gets even more complex if one considers that the majority of today's scientific fields derive from the liberal arts.

Much has been written concerning the genesis of the university, although a basic source remains Hastings Rashdall's (1858–1924) pioneering work (1895). Right from the start, then, the university developed under complex circumstances: it was subject to interactions of various factors, at time contradictory, including vocational training and the related corporatism, the liberal arts, local politics, and, above all, religion; all these left traces and confusion in modern universities. Let me here only draw attention to the complexity of this process, beginning by considering corporatism, which, in the Middle Ages, modeled the university's juridical condition.

Guild-like organizations may be beneficial, as they may degenerate into conservative actors

Acquiring and possessing goods involves struggle and competition; the rarer the goods, the harsher the struggle. A natural way to deal with the struggle—for human beings as well as in the animal world—is to join forces, maximize the yield, and then divvy up the pie. Actually, the term *university* (in Latin, *universitas* means a collectivity united by a shared interest) denotes an association that safeguards the interests of its members and occasionally monopolizes the field, or guild (although the last term was at that time used in northern Europe). Only later was the meaning restricted to the present-day definition of an institution of higher learning. Thus, to understand some aspects of modern-day universities, one must consider the concept of the guild-like organization, its merits, and its drawbacks.

Guild-like organizations and, more in general, associations can take different forms and have a long history that goes back to the dawn of time; they are repeatedly present in one form or another in ancient sources. Some are related to higher education. As far as the biblical time is concerned, Wellhausen, speaking of the period of the Second Temple, remarks that in general, "The postexilic period was inclined to form guilds and societies, as becomes evident from the genealogical and statistical lists in Chronicles, Ezra, and Nehemiah, and the scribes had every incentive to follow the general tendency" (Wellhausen 2001, 8). Goitein (1971) agrees, when he considers one of the terms to denote the Palestinian *yeshiva—chavura—* (see chapter 4). Lynch (1972, 125) notes that the Greek schools were organized along collective, rather than corporate, lines, although in the classical world one does encounter guild-like aggregations, such as that of physicians. The Romans did not develop a generalized concept of juristic personality in the sense of an entity that had rights and duties, and they had no terms for a guild-like association or a legal person, "but they did endow certain aggregations of persons with particular powers and capacities, and the underlying legal notion hovered between corporate powers, as understood in modern law, and powers enjoyed collectively by a group of individuals. The source of such collective powers, however, was always an act of state."[1] Rome saw the rise of many different types of associations (such as priestly *collegia*), which were officially recognized and privileged unions dedicated to worship. Plutarch (c. 45–120 AD) tells that the second king of Rome of Sabine origin, Numa Pompilius (r. 715–673 BC), of whom apart from some legends little is known, encouraged unions of individuals exercising the same profession in order to bring together the two ethnicities of Rome, the Romans and the Sabines. Writing between the 1st and 2nd centuries AD, Plutarch narrated what had presumably taken place more than six hundred years earlier. We do not know if it actually occurred, nor if the idea—brilliant and eventually successful—was the brainchild of Numa or someone else, or if it was only a figment of the imagination. Even if a legend, it still indicates how beneficial aggregations were considered in antiquity. Durkheim would probably classify their creation as a qualitative leap from a society held together by a "mechanical solidarity," that is, by the mere instinct of the clan, toward an "organic solidarity," that is, a more rational collaboration that better exploits the diversities of its members. This was, undoubtedly, also the early spirit of the university. Pace Durkheim, even more evolved societies ended up constructing barriers and taking on a tribal form, as was already implicit in Plutarch's text: in order to keep the artisans united, Numa "appointed social gatherings and public assemblies and rites of worship befitting each body,"[2] and worship related to profession is typical of a "mechanical solidarity."

Aggregations or guild-like associations may be beneficial but can become harmful: in Popperian terms, they are or can become closed societies. In Imperial Rome, *collegia*—beginning with clerical colleges of pontiffs—which initially safeguarded the interests of their members by offering, among other, mutual assistance, degenerated into political associations and were later in part prohibited. Only a few of public utility, such as the blacksmiths, indispensable for arms making, survived. Medieval guild-like organizations in the West later adopted various forms and names, according to their geographical location: "guilds" (from *gold*, indicating the gold that was deposited as a shared fund) for the merchants of northern Europe; *confréries* or *métiers* in France; *gremios* in Spain; *arti, compagnie*, or *fragile* in Italy; and *Zünfte* in Germany. Today, too, one finds many types of professional associations, active under names such as "union," "corporation," "lobby," "cartel," "order," "register," or even only "association" and "society." Some are genuinely dedicated to improving the working and living conditions of their members. But as well attested already by many modern thinkers, starting with Adam Smith (1723–1790) in the 18th century, not merely an exceptional few in the long run thwart competition and progress. Even more so, in a relatively free and open modern society, corporative associations can become counter-productive. A classic contemporary example is America's lobby of medical doctors and health insurances, which resists every attempt to introduce health care for all Americans. Or in Italy, during the "First Republic" (1948–1994), the communist party used (or, better said, misused) the trade unions as a lever to impose itself by paralyzing the country with strikes. Paradoxically, in the modern world, it has become difficult to dismantle these associations, and the main purpose of some, of course not openly declared, is to keep non-members out. This is part of the "paradox of freedom," dealt with by Popper on various occasions in *The Open Society and its Enemies*. Freedom with too little control can become self-defeating and give free bullies more opportunities to enslave the meek.

The medieval society into which the university (in the current sense of an institution of higher learning) was born was a closed society, both literally and in the philosophic sense of a tribal society. In the medieval Western world, every important social initiative had to be legitimated by the pope or by the emperor and could be carried out only through guild-like associations. These became particularly instrumental with the medieval emergence of cities gathering different types of tradesmen. The members of these associations were required to scrupulously respect an oath of solidarity. The "universities" (in the sense of guilds) governed themselves by means of statutes, councils, administrators, and magistrates. They imposed their own rules of conduct, often captious, taxes, and much more. They acted in the professional interest of their members, guaranteeing them the social support they needed to survive. To keep their monopoly or any specific art, they also obliged their members to keep trade secrets. All of this, naturally, came at a high price and punishment for those who failed to tow the line. One fundamental privilege of guild-like organizations was training the new recruits and conferring the artisan's license, which permitted holders to exercise the trade, and the teacher's license, which permitted them to teach the trade under the title of "Master." Nobody could exercise a trade without the guild's consensus, and this occasionally encouraged corruption. As early as the flower of cathedral schools, when it became profitable to teach students willing to pay, the number of masters multiplied, making the office of master of schools or chancellor lucrative, and scholastic simony flourished (Post 1967, 256–257).

The medieval university was born and developed in this context, between the irrational herd instinct and rational collaboration, and corporatism inherited from the Middle Ages is still present in a concealed form in modern universities. Today, currents led by academic "opinion leaders" at times create intellectual fads that aim more at personal or social success than at intellectual progress, attracting swarms of more or less organized followers, and disregarding or even silencing criticism and dissent. Moreover, notwithstanding progress and the fundamental need of science for free criticism and interdisciplinary collaboration, the paradigmatic Aristotelian segmentation of knowledge still reigns undisturbed. Those who belong to a certain field or to a certain current of thought and yet dare to express thoughts that are pertinent to different fields or currents are often ostracized. The result is studying for career, titles, and social advancement rather than for genuine love for the subject studied. The roots go back to the medieval university. Let me first outline how the medieval university came to be.

*The university was created in the medieval city from nuclei of people
studying and teaching medicine, law, theology, and the arts who were
subject to tensions of various kinds*

We still know relatively little about the genesis of the institution we call
"university." Toward the end of the first millennium, contact with the Orient brought to Europe new knowledge, including profane learning. The
European teaching system, composed primarily of religious schools in convents or associated with cathedrals, was not able to absorb all this new
knowledge. Attempts were made to found more open centers of study and
teaching. In the 10th century, or perhaps even earlier, Salerno in the south
of Italy, on the fringe of the Arab world, was already a renowned center for
medical care open to foreign learning. In the 11th century medical manuscripts were copied out at the nearby abbey of Montecassino, and there are
traces of medical literature coming from there as well, gaining momentum
in the following century and creating the basis for a methodical, theoretical,
medical instruction. We nevertheless know little or nothing concerning the
institutional aspects of Salerno as a school of medicine; teaching may have
taken place in small groups, and there is no evidence of any medical degree
being granted. In this context, Paul Oscar Kristeller (1905–1999) makes
note of an interesting detail:

> In classical Latin, the doctor is regularly called *medicus*, whereas the
> Greek term *physicus* is reserved to the student of natural science or nat-
> ural philosophy [. . .] It was not until the early twelfth century that the
> term *physicus* began to be used for the medical doctor, and it became
> gradually more frequent and remained in use up to the end of the Mid-
> dle Ages.
>
> (Kristeller 1945, 159)

Kristeller suggests that the use of the term *physicus* entailed the strength-
ening of the theoretical basis of medicine and the extension of the curriculum to other fields. Indeed, the earliest Latin version of Ptolemy's *Almagest*
was translated from the Greek by a student at Salerno (Kristeller 1945,
160). There is, however, no evidence Salerno granted medical degrees till the
13th century. The first official regulations concerning the school, including
examinations, came relatively late and are included among the constitutions
of Emperor Frederick II (1194–1250) in 1231.

The economic reawakening of the emergence of towns and the growth of
knowledge led to an increased demand for skills and, as a result, for more
study. The development of commercial activities in public administrations,
including guilds, and the struggle between the papacy and the empire over
the investiture of the bishops, required legal experts. Because both the religious and the secular authorities were reluctant to introduce profane teaching,
city students and teachers began to join forces to form universities, or rather,

independent guild-like organizations whose goal was to legitimize and support the teaching of certain subjects, as well as to acquire other rights. The process was long, gradual, and complex, and those who studied it had to struggle with fundamental difficulties of a historiographical, philosophical, and sociological nature.

The historiographical difficulty lies in the dearth of sources. Sociological and philosophical difficulties arise from the question of whether a particular institution could be considered a "university" if it had not yet received official recognition. For example, how should the two centuries of activity that preceded the official regulations of the Salerno medical school be considered? What weight does an activity or a current of thought have if it is not yet officially recognized? The answer depends to a great degree on a broader question: whether an institution is the product of a social need or vice versa. Specialized literature dealing with the topic notes a scarcity of sources and, occasionally, also makes reference to the socio-philosophical difficulty. However, less attention is paid to this latter question, and in the end many details are offered that are at times contradictory and can confuse the reader.

The following renders no claim of bringing order and precision to the historical account. I only wish to underscore the complexity of the process, subject to various kinds of tension. In various cities, nuclei were formed in teaching of the arts, medicine, law, and theology, endeavoring to become a *universitas magistrorum* (teachers' association), *universitas scholarium* (students' association, where students employed the teachers), or a *universitas magistrorum et scholarium* (teachers' and students' associations); their goal was to lay the framework for teaching, called *studium*. These associations grew slowly, surrounded by tension, not least of which that between religious and secular factions; but eventually they were recognized as "universities" by the pope, the emperor, or other local authorities. Two centers of study are considered the archetypes of the university: Bologna and Paris.

At the end of the 11th century, Bologna was a focal point of conflict between the pope and the emperor, and a number of teachers of *trivium* began to apply themselves to the study of law, above all to the exegesis of Roman law. As Olaf Pedersen (1920–1997) remarks (Pedersen 1997, 126), "In accordance with old Roman usage, law was viewed as part of rhetoric, and could be formally incorporated into *artes liberales*," being an awareness of the utility of interdisciplinarity at this early stage. Around 1080 a manuscript of the 7th century of Justinian's *Corpus Juris Civilis* reached Bologna and was examined by local scholars. The pioneer in this enterprise was one Irnerius (c. 1050–1130). Afterwards, Roman law came to play a much greater role both in teaching and in politics.

When was a "university," in the sense of a guild, established in Bologna? According to Charles Haskins (1870–1937), "The students of Bologna organized such a university first as a means of protection against townspeople, for the price of rooms and necessaries rose rapidly with the crowd of new tenants and consumers, and the individual student was helpless against such profiteering" (Haskins 1975, 9). Historian and archeologist Corrado Ricci (1858–1934), in his *I primordi dello Studio di Bologna* (1887), points out how difficult it is to ascertain the date of the foundation of the university. Yet poet and (later) Nobel laureate Giosuè Carducci (1835–1907), who taught at the University Bologna, suggested to celebrate the 800th anniversary in 1888, implying that 1088 was the founding year. The celebration took place in the presence of the king of Italy, Umberto I (r. 1878–1900), and Bologna became, rightly or wrongly, the first European university (Tosi 1989, 53–111). Around 1140, Franciscus Gratianus or Gratian, probably a Camuldulensian monk (the Camuldulensian monks are part of the Benedictine family of monastic communities; their name is derived from the hermitage of Camaldoli, on the mountains of central Italy, near Arezzo in Tuscany) and the bishop of Chiusi (also in Tuscany), systematized the many ecclesiastical laws that were produced through the centuries in *Concordia Discordantium Canonum* (Harmony between Discordant Rules of Laws), also known as *Decretum Gratiani* or *Decreta*, conferring canon law methods and principles of the Roman law. Gratian also taught in Bologna monasteries. In 1158, a number of jurists from Bologna convinced Emperor Frederick I (Barbarossa, 1122–1190) to promulgate a constitution, *Authentica Habita*, which protected itinerant students from interference by any local authorities and, in particular, prohibited their being taken hostage for debts accrued by their fellow citizens. *Habita* also established that every school was required to constitute a *societas* presided over by a *dominus* and that students were under the jurisdiction of their teacher or of the bishop, suggesting that some kind of guild already existed in Bologna before 1158. Be that as it may, *Habita* is important because it became the basis for the future imperial universities. It was an innovative step toward academic freedom, so vital for scholars to think and create. It also indicates the extent to which study had passed beyond the local borders and had become linked to the church and religion.

Universities became increasingly ecclesiastic institutions

In 1179 the Third Lateran Council took place ("Lateran" is a place in Rome) under Pope Alexander III (r. 1159–1181), ending the quarrel between the pope and the emperor and a painful schism within the church; it proclaimed, inter alia, the principle that schooling should be free of charge, a sign of downright openness. Canon 18 states the following:

> In order that the opportunity of learning to read and progress in study is not withdrawn from poor children who cannot be helped by the support

of their parents, in every cathedral church a master is to be assigned some proper benefice so that he may teach the clerics of that church and the poor scholars. Thus the needs of the teacher are to be supplied and the way to knowledge opened for learners. In other churches and monasteries too, if anything in times past has been assigned in them for this purpose, it should be restored. Let no one demand any money for a license to teach, or under cover of some custom seek anything from teachers, or forbid anyone to teach who is suitable and has sought a license. Whoever presumes to act against this decree is to be deprived of ecclesiastical benefice. Indeed, it seems only right that in the church of God a person should not have the fruit of his labor if through self-seeking he strives to prevent the progress of the churches by selling the license to teach.[3]

This meant that, as early as in the Middle Ages, everybody in the Catholic world, even the poor, at least as far as the church was concerned, had the right to an education (even if in actuality, very few got it), a principle that, still today, many universities, particularly in Europe, continue to apply. The church bore the expense of paying teachers, thus also making them dependent upon the church and extending its influence. The last sentence of the canon, implicitly admitting simony, declared meritocracy within the learning establishment.

The dependence of universities from the church gradually increased. In 1219, Pope Honorius III (r. 1216–1227) prescribed that no one in Bologna should receive the *licentia ubique docendi*—the license to teach anywhere—without the consent of the local Archdeacon, thus intensifying the link between higher learning and religious authorities. Honorius also prohibited churchmen to study medicine, implying that some teaching of medicine already existed (Sorbelli 1940, 108). Nevertheless Bologna remained, till the middle of the 14th century, primarily a secular higher learning institution. In the 13th century Bologna's *studium* consisted initially of two universities run by students (the rector was a student!), a *universitas scholarium citramontanum*, gathering Italian students, and a *universitas scholarium ultramontanum*, gathering foreign ones. The teachers, too, formed two colleges

> Requiring for admission thereto certain qualifications which were ascertained by examination, so that no student could enter save by the guild's consent. And, inasmuch as ability to teach a subject is a good test of knowing it, the student came to seek the professor's license as a certificate of attainment, regardless of his future career.
>
> (Haskins 1975, 11)

At the turn of the century Bologna was a conglomerate of universities: one for law (*universitas juristarum*) and another for the (liberal) arts (*universitas artistarum*), granting respectively the titles of *doctor* and *magister* (Sorbelli 1940, 105–106); each of these universities was divided in two *nations*, respectively, for Italian and foreign students. Slowly, the teaching of arts branched into several fields of learning: Medicine; *Humanae litterae,* the study of classical texts; philosophy; and *ars dictandi,* the art of composition of documents. The teaching of theology was introduced in Bologna as late as 1364 (Sorbelli 1940, 107 and 109).

Unlike Bologna, Paris was, from the very beginning, more oriented toward religious studies. The genesis of the university followed a different path: that of a *universitas magistrorum.* In the 12th century, Paris became a center of learning. Many schools teaching theology, the liberal arts, law, and medicine were created in a disordered manner in the city or its surroundings (Verger 1999, 25–27). The university emerged sometime during the second half of the century from a movement of teachers of liberal arts from the île de la cité to the left bank of the river, where it remained till the present day. Teachers taught initially by virtue of the license (*magister* or master) conferred to them by the chancellor of the cathedral of Notre Dame.

In 1180 one found in Paris the first known medieval Western college: called *domus pauperus,* a hostel for poor people, it was founded in 1180 in Paris by a pilgrim who returned from Jerusalem (Haskins 1975, 18; Makdisi 1981, 225–226); it was later called *Collège des dix-huit*—the college of eighteen poor students. These philanthropic colleges, reminiscent of the Islamic *madrasa*s, should not be confused with the teachers' colleges mentioned earlier in this chapter.

In 1200, Philip Augustus II, King of France (r. 1180–1223), bestowed upon the teachers and their students the privilege of being judged by the ecclesiastical authorities, giving even more momentum to the process of submission of universities to the church. The earliest pontifical recognition as a "university" occurred around 1211, and soon in 1219 the teaching of civil law was forbidden in Paris (Haskins 1975, 16). In the same year, the term *faculty,* describing a structure teaching a specific subject, appeared for the first time. By 1231 Paris had four faculties: the arts, medicine, theology, and canon law, the latter, again, in analogy to Jewish and Islamic medieval higher learning institutions. Paris, we shall soon see, was subject to tensions owing to religious pressure.

In 1257 the theologian Robert de Sorbon (1201–1274), chaplain of King Louis IX (r. 1226–1270), founded the Sorbonne College, initially for twenty poor theology students. The Sorbonne became a place of instruction and the core of the modern University of Paris. In 1304 Queen Joan I of Navarre (1273–1305), wife of King Philip IV of France (r. 1285–1314), supported the foundation *the Collège de Navarre* for students of grammar, logic, and theology, a rival college to the Sorbonne. I mention these colleges because of their relevance further on.

The church remained the institutional point of reference for knowledge throughout the Middle Ages and beyond, and it soon took control of the universities, offering them support, privileges, and protection. This process was favored by the Dominicans and the Franciscans, inter alia, because the monks were not enthusiastic to see competing "universities." Many Western universities remained confessional—either catholic or, later, protestant—until the 19th century, and their members, religious or lay, remained answerable to clerical jurisdiction. Thus, paradoxically, an institution born in part to encourage, among other, the expansion of profane knowledge became a center of conservative religious learning, subsumed into the apparatus of the church. This is one of the roots of the historical tension that still lies within the university: to quote a leading historian, the late Jacques Le Goff (1924–2014),

> It was essentially an ecclesiastical corporation. Even if all its members had not received orders, even if it counted among its ranks an increasing number of pure laymen, academics were essentially all clerks, answerable to ecclesiastical authorities, and primarily to Rome. Born of a movement heading toward secularity, they were of the Church, even when they sought to leave it institutionally.
>
> (Le Goff 1993, 72)

Universities not only replaced cathedral schools but also became institutions dedicated primarily to forming new generations of clergy and providing for the stability of the religious establishment through the transmission of the dogmas of Christianity and its culture. The liberal arts, law, theology, and, occasionally, medicine were influenced. Law faculties taught canonical law alongside civil law, and the faculty of theology became the most important. Religion is based on dogma, and dogma is the enemy of science because it blocks development and cultivates stagnation. This is one of the main reasons behind the failure of the medieval university, with a few exceptions, to foster the sciences. Moreover, dogmatic learning, even if no longer religious, remains an obstacle to the modern university, where science is often taught as though it were the ultimate truth. Science is not truth, but the ongoing quest for truth. Science is never certain, at very best ever *less wrong*.

Paris is paradigmatic as a religious university: from the middle of the 13th century, the Dominicans and Franciscan friars studying at the university requested privileges that challenged the university rule: they refused to take liberal art courses and associated only with the faculty of theology, and the controversy led to the intervention of the Pope Alexander IV (r. 1254–1261). The year 1277 saw a conflict over the scope of philosophy and its relation to theology, leading the bishop of Paris to condemn 219 propositions said to be taught in the faculty of arts of the university.

Old universities gave rise to new ones: for example, the medical school in Montpellier, emerging in the first half of the 12th century, may have been an offshoot of Salerno, or, as Rashdall (2012, 116) remarks, "The new impulse may be ascribed more directly to contact with the flourishing Jewish and Arabic schools of Spain in the twelfth century, or with the Jewish schools of Medicine in Arles and Narbonne." Regrettably we know little or nothing concerning these early medical schools. The University of Bologna gave rise to those of Reggio in Emilia, Vicenza, Arezzo, probably Padua, and to the law school in Montpellier; the university of Paris nurtured Oxford, already a center of studies in the 11th century. And according to Rashdall, Oxford grew thanks to the migration of students and teachers from the continent. A. B. Cobban (1975, 97–101), however, disagrees and argues that Oxford, like Bologna and Paris, grew gradually following a variety of factors. Oxford was, in any case, ecclesiastic from the very beginning, and because it did not have a cathedral and consequently did not have a cathedral school, the chancellor of the University was appointed by the bishop of Lincoln (Kibre & Siraisi 1978, 126). Offshoots from Oxford founded the University of Cambridge. Around 1264, Walter de Merton (c. 1205–1277), chancellor to Henry III of England (r. 1216–1272) and later to Edward I (r. 1272–1307), founded the college that took his name: Merton College. Like the Sorbonne, it became a center of learning.

To be sure universities could be created by papal bulls, by imperial decree, or by decree of local authorities. The towns that featured universities were only relatively independent: they had to answer either to the church (the pope) or to the emperor. Hence universities could rely on one of these two political forces. Bologna, for example, was a Guelph city (the Guelphs were the pope's supporter; the Ghibellines supported the emperor). From the 13th century on, dozens of universities were created, each with its own system, its own didactic methods, and its own history. German universities, all founded after the 13th century, imitated Paris. In 1386, the Palatine Elector Rupert I (r. 1309–1390), in founding the University of Heidelberg, required that it "Shall be ruled, disposed, and regulated according to the modes and matters accustomed to be observed in the University of Paris" (Haskins 1975, 20).

Beyond religion, learning becomes increasingly an instrument of social climbing

Rulers soon realized the importance of universities and began founding universities directly. For example, in 1224, Emperor Frederick II (1194–1250) founded the University of Naples seven years before the extant and nearby school of medicine in Salerno came officially to being. Naples, as well as some Spanish universities, was a royal institution for training bureaucrats. The teachers were royal employees and the sovereign conferred the degrees. In Spain, the *Siete Partidas*, the legal code composed by King Alfonso X of Castille (r. 1252–1284) between 1258 and 1265, gave to university teachers the status of knights (Pryds 2000, 93), thus officially extending their position beyond teaching to a more general social status with a status symbol. Indeed, university scholars differed from other members of crafts or guilds by being considered as (quasi) nobles, taking advantage of university studies as a means of social climbing, which continues to this day.

Scholars, indeed, constituted a new privileged class in society, the intellectuals; they were the new nobility, as suggested by contemporary references to the "new chivalry," distinguished from the feudal aristocracy since they boasted no military prowess nor material wealth, although they borrowed feudal terminology and ceremony. The scholar on the first rung of the ladder of the academic hierarchy was named "bachelor," a term originally signifying a squire.

(Kibre & Siraisi 1978, 121)

The very nature of the university as a guild-like association turned it into an institution of social climbing. Le Goff (1993, 126) remarks that "Knowledge became a possession and a treasure, an instrument of power and no longer a disinterested end in itself," quoting a 13th century text that states, "Masters do not teach to be useful, but to be called *Rabbi*." The reference is not so much to the standard Jewish title "rabbi" as to that given by the Gospels to Jesus: the academics endeavored to become lords. The title "bachelor," normally granted to a liberal art student after two years of studies, is probably derived from "bas chevalier," in French "low knight": "To transform themselves into an aristocracy, academics adopted one of the usual means employed by groups and individual wanting to rub shoulders with nobility" (Le Goff 1993, 124).

They turned their dress and the emblems of their function into symbols of nobility. The chair which was increasingly placed on a dais of seigneurial grandeur, separated them, placed them higher up, magnified them. The golden ring and the beret, which they were given on the day of the *conventus publicus* and the *inceptio*, were less insignias of a function, and increasingly emblems of prestige.

(Le Goff 1993, 125)

And, as William Eamon (1991, 334) remarks, the European medieval scientific community, closed to all but academics, became to a large extent isolated from the rest of society.

Universities differed considerably from one another

Beyond religion with its dogmas and corporatism with its careerism, the structure of universities and of their components, such as the faculties, varied considerably from one university to another. Only few medieval universities featured all four faculties: the liberal arts, medicine, law, and theology (e.g., Paris and Oxford). There were simple *studia*, and larger, more important, *studia generalia*. According to Rashdall (1936, 1: 6–17), broadly speaking, a *studium generale* had to have a number of characteristics, such as the ability to attract students from faraway countries and teach at least one of the three fields: theology, law, or medicine. A teacher of a *studium generale* was recognized as a teacher by lower-level study centers. The granting of *licentia ubique docendi* was strictly related to a *studium generale*. In the latter half of the 13th century the foundation of a *studium generale* came to be a prerogative of the pope or the emperor. The elected head of the university in the Middle Ages, as still today, was the rector or *rector scholarium*, the counterpart to the *rector societatum*, who headed the guilds. Initially, the rector was also the supreme magistrate of the university. Within or alongside the structure professors and students could belong to various groups. There were the aforementioned "nations," which gathered and safeguarded the students according to their provenance, above all those of the arts faculties who, like Islamic students, often wandered from one university to another. There were colleges of various types, generally speaking, groups of individuals who coexisted under shared rule and became established units of academic life at many universities. The level of study, too, varied.

Among the teachings in medieval universities, the liberal arts were often propaedeutic, and the faculty of arts attracted the highest number of students. Sometimes they taught subjects that today are considered elementary, such as writing, and students enrolled at a younger age than they do today. Moreover, during the Middle Ages, as indeed later, many students abandoned the university after having attended this faculty, an indication of the social, rather than the professional or intellectual, significance the university had acquired.

The classical liberal arts at a more advanced level no doubt enriched Western knowledge, as well as rediscovery of ancient texts, or, occasionally, new developments and findings. The 12th and 13th centuries saw the introduction of new works by Aristotle and fundamentals of Greco-Arab science. Albertus Magnus (c. 1200–1280) and his pupil Thomas Aquinas (1225–1274), who taught in several European universities, reconciled in their *summae* ("sum up" of knowledge—a medieval textbook) Christianity with Aristotelian philosophy into a relatively rigid learning. The development of knowledge, including that knowledge at the roots of modern science, was thus subject to religion in addition to corporatism. More in general, there was little or no awareness of the growth of knowledge, which was conceived as static and complete.

Madkdisi (1981), as seen, points out many similarities between *madrasas* and European medieval universities, in particular university colleges: structures, methods of learning, license to teach, and so forth. I would add the similarity between Jewish *yeshivot* and universities. Of course, there were many types of *yeshivot*, *madrasas*, and European colleges, and there is no evidence of direct influence. Yet, to paraphrase Makdisi (1981, 286), it is inconceivable that three cultures developed side by side for centuries without being aware of developments on either side.

Over time, medieval *studia*, with the teaching of the liberal arts, law, medicine, and theology, embraced virtually the totality of higher knowledge in the Middle Ages. The universities came to monopolize a large part of the knowledge, which was still circumscribed, in a society in which the majority of the population was illiterate.

Only medicine could claim relative independence from religion, yet all the teachings, including medicine, were basically dogmatic. For a long time, medicine remained mainly a theoretical subject, separate from surgery, a practice that was considered less prestigious and was taught outside the university because it was a "trade" belonging to a lower social level. The "surgeon" also carried out other activities dealing with the body, such as cutting hair (in Greek, *cheirourghia* means manual operation). In short, surgeons were barbers. Bologna was one of the first to study the human body through dissection (1315), a practice that, at the time, was challenged and obstructed by the church. Other universities, such as Montpellier, soon followed suit, but the bigoted university of Paris tarried.

English universities were less subject to theological and political controversies than some universities in the continent. Oxford became a center of absorption of Aristotelian philosophy and, as some historians of science claim, a cradle of modern science. A key figure in this development was Robert Grosseteste (c. 1175–1253), whom Alistair Crombie (1915–1996) considers "the real founder of the tradition of scientific thought in medieval Oxford" (Crombie 1969, 1: 27). Grosseteste was a scholastic philosopher, theologian, and Bishop of Lincoln. He argued that scientific reasoning goes through "resolution and composition": induction of laws from observations, and then prediction of particulars. Yet there is little or no evidence that he influenced early modern scientists such as Galileo, nor that they applied Grosseteste's method. At Oxford, however, the "calculators," a school of thought associated mainly with Merton College, devised a number of quantitative methods for the logical examinations and description of qualitative change and motion. One remarkable result was the "Bradwardine's rule," named after the theologian Thomas Bradwardine (c. 1295–1349), later archbishop of Canterbury: it relates the speed of a moving body to motive powers and resistances. Another achievement was the "mean-speed theorem," attributed to William Heytesbury (c. 1313–1372/1373): the overall speed of a uniformly accelerated motion to equal its means. The quantitative treatment of motion begun at Oxford was extended to Paris: Jean Buridan (c. 1300–1361), rector of the university of Paris between 1328 and 1340, rebutted Aristotle's theory that a projectile is propelled by the medium through it travels, and suggested alternatively the theory of *impetus*, that is, that motion imparted to a body is maintained: a first step towards the modern concept of momentum. Nicole Oresme (c. 1320/1325–1382), who studied at the College of Navarre, worked with Buridan, was counselor of King Charles V of France (r. 1364–1380), and became Bishop of Lisieux, devised a graphical representation of speed as a varying quality. At times, Catholic historians, endeavoring to emphasize the medieval contribution to science, produced these works as precursory to Galileo's physics, although there is no direct evidence. In general, medieval studies articulated around—and defended—dogmas, albeit in an increasingly pluralistic context.

During the centuries following the foundation of the university, the flow of knowledge handed down from the ancient Greeks and conserved by the Arabs became more intense, leading to cultural, philosophical, artistic, and scientific innovations that could hardly be accepted by an institution as anchored in dogma and conservativism as the university was. The tension between tradition and innovation re-emerged, and with few exceptions, tradition in universities prevailed. The teaching of liberal arts is important in this context because it is the cradle of modern science. The historical process is long and complex. Fields that today bear such widely used names as "physics," "cosmology," and "natural sciences," however, remained completely different subjects in the Middle Ages. They described qualitative disciplines and were a part of philosophy. All this remained within a closed intellectual circle, within the limits of accepted Christian theology and (Aristotelian) philosophy, with little opening to any new types of knowledge that could question what was considered dogmatically true.

University teaching was based on text, but the approach remained basically that of an oral culture sanctifying the text

Without entering in the details of the curricula, university teaching was all over based on texts. Reading, like writing, was then a much more complex undertaking than it is today. During lessons (*lectio*—"reading") a text was read and given purport. Before the invention of the printing press, texts were rare and precious, and the ability to memorize was still much more developed than it is today, also thanks to chanting and mnemonic techniques, which have since been largely forgotten. Texts could be read out loud at dictation speed so that students could write them down (one can imagine the level of boredom these lessons could reach). Access to books was accorded to groups, even outside the classroom; one student would read out loud to his companions. Oxford's New College, founded in 1379, introduced tutorials: older fellows supplied the education of students and drew "in return an additional allowance from the college funds besides what they may have received from individual pupils" (Rashdall 1895, 3: 216). When alone, the students would silently move their lips, articulating word for word as though reciting the passages. This was in part made necessary because spelling was not yet officially established and punctuation was almost nonexistent.

Accordingly, teachers were often *recitatores*, although some were *interpretes*, and developed new trends in scholasticism. Thomas Aquinas, who taught, inter alia, in Paris, may have fit in the first group: he used to present first a text generally and contextually, and then divided it in short extracts, devoting a lesson to each. Buridan was more speculative and applied the method of *quaestio*: posing a question to stimulate the debate with interesting results.

Although excessive reliance on a textbook, then as now, is a conservative factor, in medieval universities it was a source of indirect progress: the valorization of writing and of the written text. In Egyptian and Mesopotamian antiquity, texts were composed and monopolized by closed clerical casts of scribes and had a limited circulation. In classical antiquity texts had more circulation, although slaves were often the ones who wrote them down. Monotheism, at least since the destruction of the Second Jerusalem Temple and the emergence of Jewish scribes and rabbis, further revalued the practice of writing texts, a process that went on in the Islamic world. In the West, monks were the ones who produced texts and maintained their aura of sacredness. Medieval universities, in adopting more or less permanent textbooks, produced transcriptions that were put on the market, giving rise to an industry of written texts, which contributed to their diffusion and transformed them from luxury objects into more common study instruments. However, intellectual communication during the Middle Ages remained primitive and limited: "As long as speech and manuscript were the only means for the diffusion of knowledge, new discoveries were disseminated slowly, with a good deal of corruption and sacrifice of content, and even then to a limited audience" (Eamon 1991, 334).

In the second half of the 15th century, after Guttenberg invented the printing press and books became an instrument of communication, the universities took little advantage of the invention, if one considers the great advantages (not just practical) it could offer. Marshall McLuhan (1911–1980), in his classic *The Gutenberg Galaxy* (1962), portrays the great changes—cultural, social, and perceptive—that accompanied this invention; a primarily auditory perception gave way to a visual perception, and an abstract culture, based on philosophical argumentation, yielded to a culture of applied technique, a preamble to the industrial age. Throughout Europe books became objects of everyday use for scholars, rendering university lectures ephemeral; at the same time, starting from Italy, profane literary and scientific works were printed and disseminated. McLuhan's thesis is often summarized as "The Medium is the Massage," that is, new technologies effect cognition (resulting in social changes). In universities, however, teaching based on the medieval cult the textbook (*summa*), and of lectures based on textbooks, carried on into modern times, with printed books substituting manuscripts. At the Italian University of Ferrara (where, inter alia, in 1503 Copernicus graduated), for example, a statute dated as late 1613 states that [*sic*] "the professors had to abide by the books assigned at the beginning of the year and to the relative program; neither digressions, treatises nor questions that strayed too far from the text or the annually ordained subject matter were permitted. [. . .] And punishment consisted in the loss of a trimester of salary" (Visconti 1950, 80). In French faculties, more or less in the same time,

> Knowledge was always communicated in the same monotonous manner.
> For the first half an hour the professor would read from his manuscript

cahier or printed text; next, for a similar length of the time he would proceed to explain and develop the argument extemporaneously; then the lesson would close with a roll-call or in in some cases a question-and-answer session.

(Brockliss 1987, 59)

The Royal Society of London, founded in the second half of the same century to become one of the cradles of science, by contrast chose as its motto *nullius in verba* ("On no one's word") to declare the members' repudiation of scholasticism based on text. We will return to the academies and to the Royal Society. Yet traces of medieval reading are still present today and hinder modern university studies; many students prepare for exams by memorizing texts, exactly as Feynman (1985) complained (see chapter 1). The textbook is considered a pillar of knowledge, and often the pillar of truth. Even the most popular philosopher of science of the 20th century, Thomas Kuhn (1922–1996), in his *The Structure of Scientific Revolutions* (1962), gave excessive importance to textbooks. Kuhn actually presents a dogmatic, authoritarian, and much debated model of the development of science. According to Kuhn, the direction science should take, and, consequently, study programs, is decided by the consensus of scientists, or rather by (irrational) common sense. I will return to Kuhn later in this book.

CONCLUDING REMARKS

To summarize the main characteristics of the university from its medieval foundation, we could say that it consists in an institution that is

- *Vocational*, in that it trains physicians, lawyers, and priests in contradiction to the classical "liberal" education;
- *Guild-like*, with all its constraints and defects, including social climbing and, eventually, corruption; and
- *Religious*, with teaching that was dogmatic yesterday, remains dogmatic today, and is thus substantially closed to innovation.

These tendencies can conflict with each other, causing confusion and tension. To absorb, cultivate, and increase new knowledge, very different institutions were needed, and the Renaissance had to rely, above all, on other institutions.

BIBLIOGRAPHICAL NOTES

Hastings Rashdall's *The Universities of Europe in the Middle Ages* (3 vols., 1895), republished several times in the 20th century, is fundamental. Yet except for the excellent chapter "What is a university?", the reader is submerged with details that render it more a reference book than a reading book.

An excellent overview of the history of the university is proffered in the 1960 edition of the *Encyclopedia Britannica* under "Universities." Another overview is presented by Pearl Kibre and Nancy Siraisi in their article "The Institutional Setting: The Universities" (1978). Kibre and Siraisi, like Rashdall, distinguish between universities north of the Alps (e.g., Paris and Oxford) and south of the Alps (e.g., Bologna and Padua). The article begins, "From the close of the twelfth century onward, science found its chief institutional home in the universities or corporate associations of scholars"—a claim that I discuss critically further in this book.

The many works that deal with the history of the university concentrate primarily on individual universities: the more generalized books are often collections of historical details. Among the works dealing specifically with the medieval university, two classic popular works by two leading 20th-century historians are most helpful: Charles Haskins's *The Rise of Universities* (1975) and Jacques Le Goff's classic *Intellectuals in the Middle Ages* (1993). The first was originally a number of lectures delivered and published in 1923. The second was published in French in 1957 and translated into English in 1993. Both are fluent and witty. The second part of Le Goff's work highlight the universities' internal contradictions, in particular those between faith and reason.

A. B. Cobban's *The Medieval Universities: Their Development and Organization* (1975) and Jacques Verger's *Les universités au Moyen Age* (1999; first edition in 1973) are the most flowing scholarly works I was able to read. The last work concentrates primarily on the University of Paris, but it offers a concise description of the genesis of the universities, their relationship with ecclesiastical and temporal power, and some of the social aspects of the medieval university.

Paul Oskar Kristeller's long article "The School of Salerno: Its Development and Its Contribution to the History of Learning" (1945) offers a clear survey of the primary and secondary sources related to the Salerno Medical School.

A History of the University in Europe, edited by Walter Rüegg and Hilde de Ridder-Symoens, is a useful four-volume collection of articles on the history and development of the European universities from the medieval origins of the institution until the present day. It has been published in several languages. Volume 1 (1992) is devoted to the medieval university.

NOTES

1. http://www.britannica.com/EBchecked/topic/507759/Roman-law/41326/Cor porations, accessed June 27, 2013.
2. Plutarch, *Life of Numa,* http://penelope.uchicago.edu/Thayer/E/Roman/Texts/ Plutarch/Lives/Numa*.html, accessed April 15, 2012.
3. http://www.legionofmarytidewater.com/faith/ECUM11.HTM, accessed June 9, 2013.

7   The Age of Innovation

Renaissance knowledge owes its diffusion to institutions more open than universities

If, already in the Middle Ages, the university was a conservative institution controlled by clerics, what, then, was the cradle of modern science? There are many conjectures, none entirely satisfactory. To be sure, many institutions fostered, in one way or another, science. In the next three chapters, we will continue to follow the university's relative extraneousness to innovation and to emerging science; we will also see how other institutions were more conducive to science, for they, too, left marks upon the present-day university.

Let us, as an introduction, begin with the Renaissance, the era of awakening not only to past glory, but to fresh innovation and the root of the renaissance of science. The French term *renaissance* ("rebirth") is a 19th-century creation, although Michelangelo's pupil and biographer, Giorgio Vasari (1511–1574), already spoke of *rinascita* in his celebrated *Lives of the Most Eminent Painters, Sculptors, and Architects,* published in Italian in 1550 and, in a second edition, in 1568. Moreover, "Renaissance" has been endowed with different meanings. In general, it delineates a period of revival of classical culture, paving the way for modernity: a complex process, promoting spontaneity, creativity, and beyond—first and foremost, innovation. Strictly speaking however, the Italian Renaissance is confined roughly to the *Quattrocento*—the 15th century, and the renaissance of science begins in the next century, reaching its climax in the 17th century.

Innovations need appropriate institutions apt to develop and propagate them; during the Renaissance and the 16th century, innovations arose in many disparate fields, requiring diverse, flexible institutions. Courts—pontifical, imperial, or those of the lower nobility—were fundamental; for one reason or another, they encouraged and paid individuals or groups to practice the arts, science, and culture in general. Perhaps of even greater importance were learned societies and academies: the latter became the launching pad of science during the 17th century. Artisans' and artists' workshops, merchants' shops, and other similar sites eventually became the cradles of new applied sciences. Hospitals sought new outlets for medicine. Botanical gardens, including the "gardens of simples" (growing healing plants), alchemists' workshops, and individuals or groups practicing the occult sciences all contributed to creating the embryos of various modern sciences, such as chemistry and pharmacology. The valuable import to science of the occult in the Renaissance later became a taboo and was to some extent ignored by historians. But not to digress—informal circles, such as intellectual salons, libraries, correspondence networks, and publishers, also contributed to the propagation, if not to the creation, of new knowledge.

Vasari explains the change of attitude towards innovation at roots of the Renaissance with the intrusion of Eastern art that influenced Cimabue (1240–1302). Yet an even greater innovator was Filippo Brunelleschi (1377–1446), who designed and constructed the cupola of the Cathedral of Florence. Brunelleschi's presumed biographer, the humanist, architect, and mathematician Antonio Manetti (1423–1497), reports (c. 1480) that Brunelleschi and Donatello (1386–1466) went to Rome to study directly and in detail the engineering architecture of ancient Roman buildings:

> They made rough drawings of almost all the buildings in Rome and in many places beyond the walls, with measurements of the widths and heights as far as they were able to ascertain [the latter] by estimation, and also the lengths, etc. In many places they had excavations made in order to see the junctures of the membering of the buildings and their type—whether square, polygonal, completely round, oval, or whatever. When possible they estimated the heights [by measuring] from base to base for the height and similarly [they estimated the heights of] the entablatures and roofs from the foundations. They drew the elevations on strips of parchment graphs with numbers and symbols which Filippo alone understood.
>
> (Manetti 1970, 52)

Whether this report is true or invented, it conveys the spirit of the Renaissance. Such a spirit is also exemplified by an anecdote in Vasari's *Lives of the Artists*: a Genoese merchant ordered from Donatello a life-size head of bronze, which was made very light because it had to be carried a long distance. When the head was finished, the merchant complained that Donatello was asking too much money for it, arguing that—because he had finished the work in a month or a little over—he would be making over half a florin a day. Donatello felt grossly insulted and

> shoved the head down on the street where it shattered into pieces and added that the merchant had shown he was more used to bargaining for beans than for bronzes. The merchant at once regretted what he had done and promised to pay twice as much if Donatello would do the head again.
>
> (Vasari 1988, 1, 180–181)

The willingness of the merchant to pay twice exemplifies a spirit of openness widespread in Renaissance society but not so easily found within the university establishment. Where did this spirit come from? The richness and complexity of Renaissance culture does not simplify the study of its institutional framework and it would be as rash to emphasize as to exclude some or other institutions from the development of Renaissance culture, art, or science.

Recently, much attention has been devoted to the contribution of courts to early modern science. Renaissance culture and science could never have flowered without the impetus provided by popes, princes, and of many other patricians of different rank—noblemen, priests, or rich commoners who, thanks to their means and learning (often acquired at universities), could afford to give their support. Leonardo da Vinci, Michelangelo, Vesalius, Tommaso Campanella, Tycho Brahe, Kepler, Galileo, and many others were encouraged and given hospitality by courts. Court patronage is a broad and complex topic that was in fashion in late 20th-century history of science and has been occasionally overemphasized (see Appendix 1). I will limit myself to discussing those institutions primarily dedicated to cultivating and teaching to the fields that evolved into science, without neglecting the role played by the courts. Beyond courts, learned societies and academies were an example of centers that furthered the renaissance of science.

Learned societies is a general term, and academies were learned societies. But "academy," too, is a vague term that has to be clarified. Like the universities, the academies differed greatly from one another, even though, initially, they were inspired, as the term indicates, by Plato's Academy. James Hankins (1991, 433–435) lists seven contemporary meanings (although there must have been more). Out of these seven I would like to consider the more common of a "regular gathering of literary men"—in a broad meaning, including all types of learned men. Initially academies were spontaneous gatherings of intellectuals sharing the same interests, typically around an illustrious scholar. Normally, academies were supported by a patron who could even be the pope or a king; once the patron or the scholar died, the academy often dissolved.

Occasionally roles overlapped: individuals acted as institutions or created learned societies that eventually evolved into officially recognized academies. Academies became universities, and universities or workshops presented as— or called themselves—academies. Today, even more than in the past, the term *academy* is synonymous with an institution offering a high level of cultural, artistic, or scientific proficiency, or simply claiming competence and prestige. It is however evident that already during the Renaissance, the newly created knowledge was relatively little cultivated by universities, which entered a long period of decline, reemerging as leading scientific institutions of science and culture only after the French Revolution. A thoroughly documented work by Paul Grendler on *The Universities of the Italian Renaissance* (2001, almost 600 pages!), delving in all possible aspects of Italian Renaissance universities, produces relatively few examples of progress towards science. Speaking of the early period 1370–1425, Grendler remarks, "The famous men of letters who composed the first, second, and third generations of humanists did not want to teach in universities. Petrarch and Boccaccio fled from university teaching as if it were the plague" (Grendler 2002, 205). Later, some humanists began to join universities, but they were exceptions.

Humanism, like *Renaissance*, is a vague term used to label secular, at times mystic, thinking and approach, in contrast to universities' religious although more rational approach. Renaissance humanists sought to cultivate, above all, the art of rhetoric and language, following the classical mold. Although belonging to a basically secular movement, many famous humanists were religious, and universities, from their part, gradually opened to humanism and innovations but lacked that spirit of encouragement that spread all around to Renaissance art and culture.

The first European literary societies appeared in the late Middle Ages in France, the Netherlands, and London (Dixhoorn & Speakman Sutch 2008). In the 15th century in Italy one finds *Solitates literatorum* (literary salons), ranging from what might be called house academies—meeting in the house of a scholar to discuss matters of mutual interest or simply to use his library—to court academies, which were more elitist gatherings where participants engaged in more-or-less organized displays of erudition and learned play (Hankins n.d., 5).

As private institutions, academies were less prone to developing the tensions generated in universities, nor were they subject to corporative constraints—whether clerical or lay—, career needs, or religious or philosophical dogmatic thought. Later, some took on a juridical, guild-like form and obtained long-lasting recognition and protection of a prince or a sovereign (and that was the beginning of their decline). They could offer lectures and, occasionally, entire courses. Academies normally did not confer titles, but not seldom, membership was (and still is) flaunted as a title; moreover, as we will see, there were cases of academies merging with universities. At the beginning of the modern era, between the 15th and the 17th centuries, hundreds of academies, dedicated to various intellectual fields, were active in Europe. Most of them were literary societies, cultivating, inter alia, the vernacular in contrast to the universities, where Latin was spoken, and some were artistic, scientific, and the like. Given the great variety, even a general outline e would run too long for this chapter; therefore let me deal with a few specific cases as examples.

It would be natural to begin with the first famous Renaissance academy: the Florentine "Platonic Academy." The literature tells and retells a glowing account of a discussion group, inspired by the homonymous Greek academy, that started already meeting around 1462 under the patronage of the rich and powerful Cosimo de' Medici the Elder (1389–1464), who for all intents and purposes acted as ruler of Florence. The group of philosophers and scholars would meet in the Villa Medici at Careggi, near

Florence, under the guidance of Marsilio Ficino (1433–1499), the son of Cosimo's personal physician. Ficino was to become the greatest contemporary Plato scholar and translator, dedicating himself to the recovery of ancient texts and, in particular, the resurgence of Platonic philosophy. Alas the story seems more legendary than true, and Hankins concludes that the "Platonic Academy" was most probably only an informal gymnasium or school attended by Ficino's pupils and occasionally frequented by older persons in the role of observers and critics (Hankins 1991, 448). Yet this type of institution, too, had a role in the development of knowledge, and Ficino's contribution to the diffusion of Platonism and to scholarship in general is far from negligible.

Not much better documented is "Bessarion's academy." Basilios Bessarion (1403–1472), of Byzantine origins, was a scholar, a patron, a bishop, a cardinal, and a church diplomat who traveled much through Italy and Europe. His Roman palazzo with its library was a meeting place for intellectuals, among them Greeks, including refugees. Bessarion supported them by commissioning transcripts of Greek manuscripts and translations into Latin. In his entourage one finds, among others, the mathematician and astronomer Johannes Müller (1436–1476) from Königsberg, better known as "Regiomontanus." Bessarion brought him from Vienna to Rome. Regiomontanus has a prominent place in the history of science above all for his compilation and transmission of knowledge. Regiomontanus collected numerous Greek manuscripts and translated many of them, including works of Ptolemy, Apollonius, and Archimedes. Can Bessarion's patronage be considered an "academy"? If yes, it confirms how multifold the concept of academy could be.

One other example of academy was that of the humanist Pomponio Leto (1425–1498), the academic name of Giulio, the natural son of a nobleman of the Salernitan princely family of Sanseverino. In the fifties of the century, in Rome, he attracted disciples and admirers, forming a learned circle that worshipped, inter alia, classic paganism—a worship obviously not greeted by the church and by Pope Paul II (r. 1464–1471). The latter had Pomponio Leto and some of his followers arrested, although they were rehabilitated by Pope Sixtus IV (r. 1471–1484), himself a patron of arts and sciences. The list of such circles, in Italy as in the rest of Europe, is long (see Chambers 1995).

Renaissance private circles contributed to the growth of knowledge but could represent a danger for local rulers or sovereigns as incubator of ideas not always in line with theirs. Putting them under control, on the other hand, could bring practical advantages and prestige. Rulers began to support academies, or they founded new learning institutions alternative to universities. The 16th century became a golden age of the academies and, as Nikolaus Pevsner (1973, 12–13) remarks, whereas "Renaissance academies were entirely unorganized, the academies of Mannerism were provided with elaborate and mostly very schematic rules."—a first step towards official recognition.

During the 16th century, rulers began to found institutions alternative to universities, and some learned societies began to take an official form and their activities gained a social and political dimension competing with the university

From the Italian Renaissance on, all over Europe institutions dedicated to cultivating higher learning began appearing. Erasmus of Rotterdam (1467?–1536), for instance, whom the *Stanford Encyclopedia of Philosophy* considers "the most famous and influential humanist of the Northern Renaissance,"[1] inspired humanist and patron Jerome de Busleyden (1470–1517) to found, in 1517 in Leuven, the *Collegium Trilingue*, dedicated to that classical learning. This type of learning was not popular at the University of Leuven; the "three languages" taught were Hebrew, Greek, and Latin. Erasmus had studied at the universities of Paris and Turin, but criticized the clergy, scholasticism, and dogmatic thinking and advocated for free will and spontaneity. He, too, distanced himself from the university, although he had taught at Cambridge and could have remained there for the rest of his life. But instead, in defense of his own freedom of thought, he preferred to roam throughout Europe. Indeed,

> Erasmus' involvement in the foundation of this new trilingual institute right in their midst was a direct challenge; indeed, not only the theologians but also many in the faculty of arts resisted its foundation and its incorporation into the university. In March 1519, one of the ablest of the Louvain theologians, Jacobus Latomus, published a book flatly denying that trilingual knowledge was necessary for a competent theologian. The book was not explicitly directed against Erasmus, but no one doubted that its intention was to refute him, his biblical prefaces, and the new Collegium trilingue as well.
>
> *Stanford Encyclopedia of Philosophy*[2]

As to the *Collegium Trilingue*, it

> Provided more than a thorough acquaintance with the up to then neglected languages. It imparted a cultural shaping of its own, a characteristic spirit, a method of thought and study that had never been heard before [. . .] It was the first to disconnect study from all practical, utilitarian aims, be they academic degrees or the admittance to special functions; its lectures were not part of a set curriculum: they only served to enrich and adorn the mind, and anyone who wished, could avail himself of them: besides being taken freely, they were also imparted freely, although given by the best qualified men.
>
> That freedom, appealing to the intellectual interest of the hearers rather than to any material advantage, made the New College eminently

modern; it was made more so by the method introduced. Instead of the text of a secular manual or a *Summa*, or of the word of the master, it accepted as foundation of knowledge and of scientific certitude only the deductions derived by personal study and research from the actual subject under observation, or from authentic, unobjectable testimonies, if it could not be brought under immediate survey.

(Vocht 1951, v–vi)

The *Collegium Trilingue* inspired King Francis I of France (r. 1515–1547) to found the *Collège des lecteurs royal*. The French "Royal Lecturers" are an early example of a state higher learning institution that attempted to free itself from the constraints of tradition, at least initially. More precisely, in 1530, the king, following the advice of the humanist Guillaume Budé (1467–1540) and of Jean du Bellay (1492–1560), bishop of Narbonne, founded a French *Collegium Trilingue*: he established two "royal lecture-ships"; one in Greek, the other in Hebrew (Lewis 1998). This was the seed of a College of lecturers to promote the teaching and the study of humanistic disciplines such as Ancient Greek, Hebrew, and Mathematics. Yet Erasmus turned down as well the offer to become a Royal Lecturer. Its pedagogy was revolutionary: it offered free courses independent from each other and open to all, without imposing exams or grades or conferring degrees. The college soon had to face opposition from the Sorbonne fearing an opening to Lutheranism. In 1533 Budé prevailed upon King Francis to refrain from prohibiting printing in France as had been advised by the Sorbonne. The number of subjects taught at the *Collège* grew, and the Lecturers—eminent scholars from all over Europe—formed a guild-like association that, in 1610, was officially called the *Collège Royal*. After changing its name several times, in 1870 it became the *Collège de France* and is still active and prestigious today. It could be taken as a paradigmatic how-to for running a modern university.

A famous 16th-century Royal Lecturer was Peter Ramus (Pierre de la Ramée, 1515–1572), a scholar who graduated at the College of Navarre. In 1543 he was forced to abandon teaching as a result of his criticism of Aristotelianism and the curriculum of the Paris University. Ramus's recommendations included a new dialectic method to substitute for the Aristotelian dialectic. He was later rehabilitated and, in 1551, nominated Royal Lecturer by Francis I's successor, Henry II (r. 1547–1559). Ramus's conversion to Calvinism and his Reformist ideas concerning knowledge and the university did not facilitate his activities, and he was forced to flee Paris several times. He was murdered in 1572 during the Saint Bartholomew's Day Massacre of the Huguenots.

In parallel, in Italy, private academies became better established. The first recorded fixed rules of an academy are those of the literary *Rozzi* academy (academy of the course-mannered ones; the names adopted by the academies much testify to the sense of humor of their member) in 1531 in Siena

(Pevsner 1973, 13), which still exists as a local academy in the same city, concerned above all with theater.

Some academies obtained official recognition and privileges; basically, they, too, became guild-like associations. As noted previously, back then such associations called themselves "universities," and what we today call "universities" were differentiated from the other guilds by the term *studium*. The confusion extends beyond merely terminology. Official recognition was advantageous for both the academics and their official patrons, at least in the beginning. The former received legitimization, financial support, privileges, and other perquisites; the latter (above and beyond an intellectual interest that was often sincere) gained prestige and, at the same time, control. The question remains: to what extent did this official recognition encourage the growth of knowledge, or rather, where did personal or social interest end and scientific interest begin? The answer is not easy and varies from case to case: the 16th-century *Accademia Fiorentina* or Florentine Academy may be a good example.

The Florentine Academy had an intriguing start. The stimulus may have come from Padua where, in 1540, one Leone Orsini (1512–1564) founded the literary *Accademia degli Infiammati* ("Academy of the Inflamed Ones"). The *Infiammati* offered lectures on vernacular poetry for only a few years but induced the emergence of the chartered and prestigious *Accademia Fiorentina*. The story is as follows: Always in 1540, a group of Florentines, mostly merchants, gathered privately, but formally, for cultural reasons, calling themselves *Gli Umidi*, which could be translated as "all wet"—to contrast the *Infiammati*. According to Armand De Gaetano, the name of this "Bohemian" academy was adopted for its connotation to fertility; the records of the academy report, "They called it the sit of humidity—a name of great etymology and no small invention."[3] As the academy grew, some its members endeavored to broaden the interest of the academy and gain the support of the Florentine duke, Cosimo I de' Medici (r. 1537–1574, from 1569 Granduke of Tuscany; not to be confused with Cosimo the Elder, who lived in the previous century). They could offer him the consolidation of the Tuscan dialect, then the language of the emerging Tuscany and the candidate to become the Italian language. Cosimo, from his part, may have feared that the *Umidi* could become incubators of political ideas not in line with his, and he "encouraged" some of them to dissolve the Academy; more explicitly, like a good "godfather," he offered them an alternative they could not refuse: to become members of a new literary academy financed and controlled by himself. Less than half a year after its foundation, the *Accademia degli Umidi* dissolved, making room, in 1541, for the "Florentine Academy."

> By making the meetings public and by endorsing its program, he could keep the leading men in Florence occupied in activities which were

nonpolitical. So he took over the Academy, assigned stipends to its officers, and tried to get as many leading scholars as he could to join it.

(De Gaetano 1968, 33)

This is only the beginning of the story. To understand the next developments, which were outcomes of genuine intellectual endeavors and literary interests as well as political intrigues, one should to go back to 1473,
when Lorenzo de' Medici *il magnifico* (1449–1492) transferred most of
the Florentine *studium* to Pisa. The move was both on financial and political grounds: it was an intelligent, although fairly unsuccessful, attempt to
empower Pisa as a leading center of study in order to counterbalance the
political power of Pisa's rival, Florence. Only a few teachers and a rector
remained behind in Florence, and students had to travel to Pisa in order
to graduate. In 1542 Duke Cosimo I decreed that "the authority, honors,
privileges, rank, salaries, emoluments, and all, attained by and belonging to
the Rector of the Studio of Florence, from this moment on belong entirely
to the Magnificent Consul of the above-mentioned Florentine Academy"
(Maylender 1926–1930, 3, 4). This promotion of the "Consul" (i.e., president) of the Florentine Academy to the rank of rector conferred on the
Academy the dignity of a university, both in the guild sense and in the modern sense of the term. As Michele Maylender (1863–1911) notes in his posthumous, encyclopedic work on Italian academies, the prerogatives of the
Consul of the Florentine Academy

> Were not limited to the sole title as General Rector of the Florentine Stu-
> dio, but [. . .] on the basis of the associative statutes and ancient custom,
> he exercised his jurisdiction over professors, doctors, scholars, servants,
> and other attendants of the University, in the lawsuits of booksellers,
> authors, and teaching staff in all things pertaining the studium. The
> Consul exercised that jurisdiction even over Academics in conjunction
> with other Law Courts of the City, whereas in disputes regarding the
> public Studium, every other tribunal and interference were excluded,
> and the Consul acted as sole judge.
>
> (Maylender 1926–1930, 3, 4–5)

Within two years, the Duke of Tuscany transformed a private academy
into a semi-university with a teaching staff and public lectures. The Duke,
however,

> Kept a close check on the subject matter of the *lezioni*; on April 27,
> 1550 he had ordered that no one was allowed to lecture either at public
> or private meetings without the approval of his *auditore* [the Tuscan
> minister of justice].
>
> (De Gaetano 1968, 34)

And,

> To keep a closer watch on its activities the Duke decreed on October 27, 1550 that the Academy meet in the Sala del Consiglio de Dugento in the Palazzo Vecchio. He sweetened the order by allowing the consul of the Umidi to participate at the meetings of the Grand Council of the Republic.
>
> (De Gaetano 1968, 35)

The Florentine Academy, despite all, was a new type of intellectual institution; if not exactly formative, it was without a doubt both popular and at a cultural high-level. To be sure, the Academy operated in a society that was still predominantly illiterate, and only a small percentage of the population was able to follow the lectures. Even as such, it was remarkably successful. It was a secular institution and thus more open. It held both private and public meetings, first at the old Palazzo Medici, later at the Palazzo Vecchio (Pevsner 1973, 14). The public flocked to the lectures held in the university's lecture hall, and up to two thousand auditors could participate. As the leading sociologist Robert K. Merton (1910–2003) notes, the interaction among many participants can enhance quality and innovation (Merton 2001). Leonardo Olschki (1885–1961) compared it to a *Volkshochschule* or a modern *Folkeuniversitet*—adult learning institutions of 19th-century Scandinavian origin, many of which are also active in Germany today (Olschki 1965b, 74). The academy also encouraged the translation of classical works into the vernacular. Previous private academies may have been more open in the philosophical sense, but socially, they were relatively closed. The Florentine Academy was more prone to political control and thus less independent, but opening to the public was, per se, an important step forward. It soon became the factor uniting and controlling Tuscany's cultured class, playing an important role in disseminating new ideas, including philosophical and scientific concepts (Helbing 1989, 48). Lecturers at the Florentine Academy included the prominent humanists and men of letter, Giambattista Gelli (1498–1563) and Cosimo's opponent, Benedetto Varchi (1503–1565); the Aristotelian philosopher and Galileo's teacher, Francesco Buonamici (1533–1603); and, in 1588, Galileo himself, at the time a 24-year-old jobless mathematician delivering a debut lecture on the size and geography of Dante's Inferno—an interesting example of interdisciplinarity.

An official institution like the Florentine Academy may have encouraged young intellectuals but, obviously, had its limitations. An indirect consequence was the foundation of the *Accademia della Crusca*: in 1582, a number of members of the Florentine Academy, who deemed the academy too formal, abandoned it, and it in 1585 they formed the *Crusca* academy. *Crusca* means "bran," and the name reflected the desire of its members to distance themselves from what they considered the pedantry of the Florentine

Academy, which they countered with *cruscate*, playful meetings with trivial speeches and frivolous conversations (bran is the broken seed coat of grain after it has been milled). In 1783 the Florentine Academy, the *Accademia della Crusca,* and a third literary academy, the *Accademia degli Apatisti,* were merged. The Florentine Academy faded away, whereas the *Crusca,* in 1811, regained independence and became the main institution of reference and for the promotion of the Italian language till the present day.

Whereas the Florentine Academy, like the *Accademia della Crusca,* straddled the spheres of academies and universities, the *Accademia del Disegno,* founded in 1563 also by Cosimo I de' Medici, on Vasari's proposal, had its roots in another Renaissance cultural tradition which contributed to the growth of knowledge: artisan's and artistic workshops. Interestingly enough, Maylender does not mention this academy, probably because he did not consider it "academic" enough despite its importance. The topic is complex, and, thus, we will deal with it in the next chapter, dedicated to Renaissance workshops.

Catherine de' Medici (1519–1589), wife of Henry II and mother of Henry III (r. on France 1574–1589, also former king of Poland), may have been instrumental in bringing the Tuscan legacy to France as far as court academies are concerned. Frances Yates's (1899–1981) detailed study concerning *The French Academies of the Sixteenth Century* (1988, first published in 1947) traces their origin in Florentine Neoplatonism and displays the wealth and complexity of French intellectual life outside universities, strained between religion and mysticism. There was, among others, the *Pléiade,* a group of Poets led by Pierre de Ronsard (1524–1585); one of its members was the prolific Jean-Antoine de Baïf (1532–1589), natural son of Lazare de Baïf, scholar and French ambassador at Venice. Jean-Antoine founded in 1570 the *Académie de Poésie et de Musique* with the purpose of reviving classical poetry and music. Court interest and entertainment led to the *Académie du Palais,* meeting in the Louvre under the Patronage of Henry III, including the Poet and Royal Lecturer, Cardinal Jacques Davie Du Perron (1556–1618).

> Several Philosophical Questions were there discussed, of which some were to be found in manuscript or in print, under the title *Discours Académiques,* and other names. With the approval of the Queen-Mother, Catherine de Médicis, and the brothers of the King, the Dukes of Alençon and Anjou, this body was the first to be officially known as an academy; continued in the Louvre under the inspiration of the poet Pibrac, the meetings and organization disappear about 1584, leaving little beyond a vague memory.
>
> (Brown 1934, 6–7)

The academy active at the palace was a forerunner to the 17th-century French academies, which are discussed in chapter 10. As far as learning and science are concerned, the founding of state academies entailed advantages. Yet the founding of scientific state academies was a gradual process. The first circles devoted to "science" (here I mean the precursory to what we call today "hard" sciences) were private.

In the middle of the 16th century, the first private "scientific" academies appeared

In the 16th century the first rudimental "scientific" societies (i.e., dealing with fields precursor to modern "hard" sciences) started to appear. William Eamon and Françoise Paheau draw attention to the *Accademia Segreta* of Girolamo Ruscelli (c. 1518–1566), active in the middle of the century in Naples. Ruscelli was an erudite, well-known humanist who had studied at the University of Padua and had joined the court of Cardinal Marini Grimani (c. 1489–1546; Eamon & Paheau 1984, 329). Little is known about Ruscelli's academy, except for an impressive description in the proem to his book, *Secreti nuovi* (1567). It relates that the academy was financed by its members, a mixture of people from several parts of Europe, including a prince, and it employed attendants, servants, apothecaries, goldsmiths, perfumers, painters, herbalists, and gardeners. Its members tested recipes and "secrets" they found in manuscripts or books through experimentation. Most of the recipes were medical; the remainder included cosmetics, alchemical processes, metallurgical recipes, and the like. The work of the Academy was kept secret. Eamon and Paheau conjecture that the secrecy was an outcome of the tense political circumstances in Naples under the Spanish domination and the activity of the Inquisition, although there is more to it: it was, in general, still common in those days to keep discoveries as trade secrets, contrary to commonly accepted practice in science today. This is also one of the reasons why we know so little about these and other secret academies.

The condition that science should be public was set in the 17th century by Robert Boyle (1627–1691—see chap. 10), who suggested that in science an experiment should be ignored unless it is repeated by independent performers and declared repeatable. The idea, however, was older, and even in Ruscelli's description there is a hint in Boyle's direction: "In doing such experiments we adopted an order and method, one better than which cannot be found or imagined, as will be recounted next. Of all those secrets which we found to be true by doing three experiments on each" (Eamon & Paheau 1984, 340).

One other Neapolitan academy, bore the same name as Ruscelli's, "*Accademia dei Segreti*" or "*Academia Secretorum Naturae*." The Academy was founded around the 1560s in Naples by Giambattista Della Porta (1535–1615), a polymath, a dramatist, and a sorcerer as well as author of *Magiae Naturalis* (1658), a work concerned, inter alia, with occult philosophy, astrology, alchemy, mathematics, meteorology, and natural philosophy. Each aspirant to join his academy was required to have discovered a secret of nature. The academy met in one of Della Porta's dwellings in the *Due Porte* section of Naples, so-named in reference to two entrances to caverns. The activities of the academy came under suspicion of sorcery, and the academy was closed by Pope Gregory XIII (r. 1572–1585) in 1578. In 1589, however, in the second edition of his *Magiae*, Della Porta reports the experiments performed by his academy (Clubb 1965, 12). Della Porta later encouraged—and joined—Prince Federico Cesi's Lincean Academy, whose most famous member was Galileo Galilei.

No doubt the relations between religion, science, and science teaching institutions had to be regulated, and a brief digression is required. A seed was planted in the middle of the 16th century during a central political and religious event: the Council of Trent (1545–1563). It initiated the Counter-Reformation, which was the church of Rome's response to the Lutheran Reformation, by closing its ranks through calls for increased discipline and piety. At its Fourth Session, on April 8, 1546, the Council hermeneutical guidelines were to become the ground for the most famous controversy between religion and science, the "Galileo case." The Council forbade an interpretation of the text as follows:

> No one relying on his own judgment shall, in matters of faith and morals pertaining to the edification of Christian doctrine, distorting the Holy Scriptures in accordance with his own conceptions, presume to interpret them contrary to that sense of holy mother Church, to whom it belongs to judge to their true sense and interpretation, has held and holds, or even contrary to the unanimous teaching of the holy Fathers.
> (Finocchiaro 1989, 12)

The guidelines were written primarily to prevent novel interpretations such as those the Protestants were inclined to make. Furthermore, as far as teaching is concerned, the Fifth Session (June 17, 1546) issued an additional decree concerning the teaching: "No one may be admitted to this office of instructor, whether such instruction be public or private, who has not been previously examined or approved by the bishop of the locality as to his life, morals and knowledge" (Finocchiaro 1989, 12). I mention the Council of Trent because it is pertinent to Galileo's campaign to separate science from religion, to which we will return, anon.

Meanwhile scientific activity outside the universities gained momentum all over Europe. In 1582, in London, the College of Physicians, which had been chartered in 1518 by King Henry VIII (r. 1509–1547) to oversee the medical profession, grant licenses, and punish abuse of power, established the annual Lumleian Lectures on surgery to diffuse the general knowledge of anatomy. The lectures were endowed by the book collector John Lumley (c. 1533–1609) and by Richard Caldwell (c. 1505–1584), one of the presidents of the College. The chosen lecturer had to hold lectures for a period of seven years. The Lumleian Lectures are still held today.

One other president of the College of Physicians was William Gilbert (1544–1603), pioneer of the study of magnetism and originator of the term *electricity*. Gilbert coined the Latin term *electricus* meaning "like amber" or "of amber" (*elektron* is the Greek word for amber, with its attractive properties). Gilbert studied medicine in Cambridge and then traveled around Europe. In Italy he met—and may have been influenced by—Della Porta. In 1573 he returned to England to work as a physician. It is said, although widely doubted, that he, too, ran an academy in his own house in Colchester, in which, with a group of friends, he discussed scientific topics and carried out experiments. In 1600 he became president of the College of Physicians. From 1601 he served as Elisabeth I's (r. 1558–1603) physician, and after her death, to her successor James I (and VI, as King of the Scots, r. England and Scotland 1603–1625). Gilbert investigated the properties of electricity and magnetism distinguishing, inter alia, the lodestone effect from static electricity produced by rubbing amber; he also noted that when magnets are cut, each part becomes a new magnet. Agreeing that the earth rotates (without, nevertheless speaking of heliocentrism), he conjectured that the center of the Earth was of iron and that the earth is a gigantic magnet. His *De Magnete* (1600) describes many of his experiments, and in the introduction, Gilbert, like Ruscelli before him, notes the importance of repeating experiments:

> Whoso desireth to make trial of the same experiments, let him handle the substances, not negligently and carelessly, but prudently, deftly, and in the proper way; nor let him (when a thing doth not succeed) igno-rantly denounce our discoveries: for nothing hath been set down in these books which hath not been explored and many times performed and repeated amongst us.[4]

Marxist historian Edgar Zilsel (1891–1944, see Zilsel 2000, 92) consid-ers Gilbert "The first academically trained scholar who dared to adopt the experimental method from the superior craftsmen and to communicate the results in a book not to helmsmen and mechanics but to the learned public." Much the same can be said on Galileo, who admired Gilbert and expressed similar thoughts as far as experiments are concerned, and on William Harvey (1578–1657).

Harvey, best known for the discovery of the circulation of the blood, joined the College of Physicians in 1607 after having studied in Cambridge and Padua. From 1615 to 1656 he was Lumleian lecturer, and it was in the course of these lectures that he presented his work on the circulation of blood, later, in 1628, in Frankfurt under the title *De Motu Cordis*. Harvey served also as physician to James I and Charles I (r. from 1625 until his execution in 1649).

One other contemporary scientific institution alternative to the university was Gresham College, founded in 1596. Sir Thomas Gresham (c. 1519–1579), a wealthy merchant, financier, diplomat, and founder of the Royal Exchange, had left funds to institute a college in which seven professors should read lectures, one each day of the week, in astronomy, geometry, "physic," law, divinity, rhetoric, and music. It is nevertheless debated to what extent Gresham College was more "open" than contemporary universities. Still, it played a role in the establishment of the Royal Society of London (see chapter 10). The great advantage of these institutions was being secular and, thus, more open to innovation than clerical universities.

Universities evolved but remained basically religious institutions with the limitations involved

The intellectual effervescence outside the university encouraged improvements within those institutions, although they were too limited to restore its monopoly on knowledge and higher teaching. Starting in the mid-1400s, Italian universities opened to humanistic studies, with Bologna leading the way. New disciplines, such as poetry, history, ethics, natural philosophy, and cosmography, were introduced into the curricula of many universities. Padua attempted to broaden the horizons of Aristotelianism. Pisa, Florence, and other universities opened to Platonic studies, and Ficino himself taught in 1466 at the University of Florence (Grendler 2002, 297). His disciple, Francesco Cattani di Diacceto (1466–1522), was one of the few professors of philosophy who continued to lecture in Florence even after most of the teaching was transferred to Pisa; he taught Aristotle, but studied Plato. Pisa itself became an important center of Platonic studies, even though, as Charles Schmitt (1933–1986) remarks, it remained the "monks' university":

> It was still quite clerical in orientation, for example, and quite a number of the staff were in orders. Indeed there must have been some foundation to the early seventeenth-century saying which characterized four universities and: "Bologna innamorati, Padova scolari, Pavia soldati, Pisa frati."[5] The curriculum was still very medieval in tone, though some influence from the humanist movement could be detected.
> (Schmitt 1972, 248)

Perhaps the university teaching that was best able to adapt itself to the times was medicine, which was less dependent on religion. Modern medicine still contains many Latin terms in keeping with the tradition of the medieval university. As mentioned previously, in the 14th century, dissection—a manual art—was gradually introduced into the university. Combining theory and practice, dissection went against the traditional great divide between the academic and the vocational, established in classical antiquity. In 1482 Pope Sixtus IV issued a bull allowing bishops to give the bodies of executed criminals and unidentified corpses to physicians and artists for dissection. As Grendler (2002, 329) remarks, Italian universities performed anatomical dissections long before northern universities: "The first known dissection of a human body at the University of Paris occurred in the 1470s and the first known public anatomy at Heidelberg, done by an Italian, only in 1574." Medical developments owe much to the university, and in this regard the University of Padua was a leader. Andreas Vesalius (1514–1564), the physician who revolutionized the study of medicine in the 16th century by shifting the center of mass from written texts to dissection, was essentially an academic.

Vesalius studied at Leuven, Paris, and Padua, where he taught. It was during his period in Padua that he made his greatest contributions to medicine. He left Padua to enter the service of Emperor Charles V (1500–1558) and had to follow the emperor in his military campaigns, an activity that left little time for study. At court, he was mocked by his colleagues because he "lowered" himself to "socially inferior" practices reserved to surgeons and deemed unworthy of doctors. In 1544, Cosimo I de' Medici offered him a chair at the University of Pisa in a sincere attempt to further the expansion of the university, but Vesalius was unable to accept due to his commitments to the emperor (O'Malley 1964, 203).

In 1594, the first anatomical theater was inaugurated in Padua. This theater, which is still perfectly preserved, consists of a dissection table standing in the center of a round room featuring six tiers with railings, all in walnut. The practice was still a delicate issue, and the corpses were hauled up from an underground canal through a trapdoor hidden beneath the dissection table. The table was constructed in such a way that, in case of a surprise inspection, it could be quickly flipped over, dropping the corpse into the canal and displaying the body of an animal that had been previously tied to the underside of the table. However, as Allen Debus (1970, 6) remarks, "The significance of Padua should not blind us to the fact that many other universities were extremely conservative and that their prime function was generally thought to be the preservation of the learning of the past rather than the quest for new knowledge through research."

While Italian universities were taking small steps to improve various professions and include humanistic studies, north of the Alps, mystic iatrochemist Jan Baptist van Helmont (1579–1644) initially refused an M.A. degree from the university of Leuven "since I knew nothing substantial, nothing true [. . .] Seeking truth and science [. . .] I withdrew from the University" (Debus 1970, 17; van Helmont did, however, return to Leuven to earn an M.D. in 1599). And, as far as Paris and Oxford are concerned, they "appeared to have reached the zenith of their importance in mathematical, natural, and physical science, and medicine, in the thirteenth and fourteenth centuries and thereafter gave evidences of decline" (Kibre & Siraisi 1978, 124). Till the beginning of the English Civil war (1642), the universities of Oxford and Cambridge

Attracted rising numbers of young gentry, along with the offspring of professional and lesser social groups. For many gentlemen, frequently staying for only a few months and rarely bothering to graduate, university was rather like a fashionable holiday camp, most of their time being spent on hunting, drinking, gaming, and feasting.

(Clark 2000, 37)

These universities nonetheless remained cradles of future administrators, politicians, and scientists. One prominent politician and philosopher of science was Francis Bacon, with whom we will deal in chapter 10. The English monarchs and the reformed church, in particular, were concerned with the universities as producers of the future leaders. In France, too, the universities' role came increasingly to that of educating civil servants (Brockliss 1987, 445).

Progress was made in strictly ecclesiastical institutions, as well, even if such institutions would, of course, not teach the Copernican theory. During the Counter-Reformation, the Jesuits created their own educational system, which included higher education and, in a certain sense, competed with traditional universities. In 1552 Pope Julius III (r. 1550–1555) authorized Jesuit colleges to confer degrees in theology and philosophy. Nonetheless, the Jesuit colleges and universities, too, remained entrenched in conservative thought. The Jesuit *Collegio Romano*, founded in 1551, was, for a short period of time, the most important scientific institution in Counter-Reformation Europe and was also involved in Galileo's discoveries. But even the mathematicians at the *Collegio Romano* were forbidden to overstep a certain limit set by theology. It took centuries to make the university an innovative institution.

CONCLUDING REMARKS

- From the Renaissance onwards, open non-clerical institutions, such as academies, were fertile ground for encouraging the growth of knowledge and developing science.
- During the 16th century academies began to be officially recognized with related advantages and disadvantages.
- The universities—guild-like, clerical, dogmatic, and conservative— were unable to accommodate and cultivate the new Renaissance knowledge and lost their monopoly on culture.

BIBLIOGRAPHICAL NOTES

It is even more difficult to write a comprehensive history of the academies, not to speak of other learned societies, than it is to write a history of the universities. As far as Italy is concerned, Michele Maylender's *Storia delle accademie d'Italia* (1926–1930) is an encyclopedic work providing precious historical information about the academies that is still useful, albeit excessively rhetorical. A short outline is presented by Eric Cochrane's "Le accademie" (1983). David Chambers's "The Earlier 'Academies' in Italy" (1995) is a short article considering critically our knowledge of early Italian academies and emphasizing how multifold the notions of academy could be, from antiquity onwards. Michel Plaisance's *L'Accademia e il suo principe* (2004), written in French despite the Italian title, is a detailed and documented history of the Florentine Academy. The mass of details, however, makes reading difficult for non-experts in contemporary Florentine cultural history. De Gaetano's "The Florentine Academy and the Advancement of Learning through the Vernacular: The Orti Rucellai and the Sacra Accademia" (1968) offers a clearer description of the founding of the academy.

The Scholarly Societies Project (http://www.scholarly-societies.org/, accessed September 13, 2013) is a useful website providing updated files with historical data of all the academies throughout the world. Pevsner's *Academies of Art. Past and Present*, published initially in 1940, beyond fine art academies, considers the general context of academies since the Renaissance. William Eamon's "From the Secrets of Nature to Public Knowledge," published initially in 1990, is an excellent outline of the sociology of secrecy from the Middle Ages to the 17th-century scientific revolution. O'Malley's *Andreas Vesalius of Brussels, 1514–1564* (1964) is an impressive biography of Vesalius.

Many historians support the thesis that the roots of modern science lie outside the university. One classic, almost encyclopedic, study is the three volumes *Geschichte der neusprachlichen wissenschaftlichen Literatur* (1919–1827), originally published in 1922, by Leonardo Olschki. Taking his cue from Renaissance scientific literature written in the vernacular, he suggests that the authors, scholars, artists, and technicians wrote in a popular, as well as innovative, manner, thus forming the roots of Galilean science, which he discusses in his third volume. Olschki's work, a posteriori, might seem superficial and biased. Like many other studies, it neglects the general illiteracy of Renaissance society. Nonetheless, it is a useful and pioneering book based on painstaking research and a good foundation for later studies of the history of Renaissance science and technology.

Edgar Zilsel, a Marxist scholar and disciple of Olschki, followed in the same historiographic tracks. His suicide in 1944 interrupted his work before its completion. A number of his articles were gathered in a posthumous volume under the title *The Social Origins of Modern Science* (2000); they stand out for their clarity and are based on both his erudition and conviction (influenced by Bacon, positivism, and Marxism) that Renaissance developments required direct contact with reality, a contact that universities were no longer able to provide. Zilsel highlights two social classes as the promoters of new science: the more elevated class of humanists composed, to a great degree, of functionaries, citizens, and noblemen; and the lower class comprising relatively prestigious professionals, such as surgeons, engineers, artists, and even simpler artisans.

NOTES

1. http://plato.stanford.edu/entries/erasmus/#PolPac, accessed August 10, 2013.
2. Ibid.
3. De Gaetano 1968, 30: "*La chiamarono il seggio degli humydi di nome veramente di grandissima Thymologia et non piccola inventione.*"
4. http://www.gutenberg.org/files/33810/33810-h/33810-h.htm, accessed January 5, 2014.
5. "Those of Bologna are in love, the Paduans are students, the Pavia ones are soldiers and the Pisan ones—monks."

8   Learning the New Techniques

Higher learning institutions have been described, through the ages, by different terms, often confusing

We have dealt with the relations and tensions between higher learning institutions and the growth of knowledge in the early modern era. Before treating the learning of applied sciences—and to avoid confusion—allow me to summarize the historical meaning of some related terms.

University, during the Middle Ages and the early modern era, was a general term to denote a guild. The first universities (in the more common contemporary usage as higher learning institutions) were corporate associations of teachers, students, or both. They were distinguished from other "universities" (i.e., guilds) by the terms *studium* or *studium generale*. Through the centuries, the term *university* was gradually restricted to denote a higher learning institution.

Much the same can be said about the genesis of *college*. In ancient Rome *collegium* had a meaning close to that of "guild," more precisely, an association with legal rights and obligations. In the Middle Ages colleges were initially residences for poor students. As Makdisi has pointed out, there are similarities between the European medieval colleges and the Islamic *madrasas* (schools, places of study). Later, the term *college* was applied, inter alia, to a group of people with special duties and powers, such as the London College of Physicians. Today it is often used as a synonym to a university, mainly for undergraduates, or—at least—of part of a university. It can also denote a secondary school with boarding.

Academy, on the other hand, originated from "Hecademus," the name of a classical mythological hero: Plato had established his school in a gymnasium located in a grove outside Athens's walls named after that hero. The term *gymnasium,* too, may be confusing because in classical time it was a facility where young men trained naked for athletic competitions (*gymno* means naked). During the Renaissance, *academy* acquired many different meanings. It was generally used to denote a regular meeting of intellectuals, or merely the place (i.e., the building) where they took place; it could, however, also denote a private humanist school more or less analogous to a modern secondary school. Or the word could be used as a synonym to university as a higher learning institution. Some of these meanings are still in use today. Today the term *academy* has come to indicate many types of educational, cultural, or scientific institutions, be they of high level or aspiring to such. In this cauldron of learning institutions at different levels, the one directly precursory to today's university distinguished itself for being a school of higher learning for clergy, even if at the time it was called an "academy." In this chapter we will deal, inter alia, with one type of academies, which may have been at the roots of modern technological, thoroughly vocational, higher learning: fine art academies. Before doing so, let me say a few words on medieval and Renaissance applied learning.

Renaissance guilds monopolized the teaching of crafts—traditionally so little-esteemed and transmitted from master to apprentice

Applied science, despite its importance, has been long downgraded by higher learning: I will not so much deal with the history of technical or technological higher learning as with the complex relationship between technical or technological higher learning and university teaching.

The late Middle Ages and the Renaissance saw an influx of new knowledge, both theoretical and practical, coming from the orient or from Spain. The resulting developments, in turn, led to new intellectual, economical, and technical needs and developments. On the practical level, there was an upsurge in fields such as architecture, metallurgy, hydraulics, the construction of machinery, the arts, commerce, economy, and other branches, all of which called for increasingly specialized training.

Most important, Hindu-Arabic numerals were introduced in the 13th century, simplifying calculations, which up to then performed using letters of the alphabet to indicate numbers. This, of course, had a considerable impact on both applied and pure mathematics. One owes this development primarily to Leonardo Fibonacci (c. 1180–1250), also known as Leonardo of Pisa, the son of a Pisan *notabile* (Pisa was then a powerful maritime republic), who traveled the Mediterranean widely and learned new arithmetical procedures from the Arabs. Fibonacci gathered and published these procedures in 1202 in his *Liber Abaci* ("The Book of Abacus")—a treatise or arithmetic written in Latin presenting Arabic numerals; positional calculations; new methods for carrying out arithmetical operations, such as the "rule of three" (i.e., cross multiplication); methods to extract the square and cube roots of numbers; and similar innovations.

In general, European society increasingly needed experts: engineers, architects, administrators, accountants, military officers, experts in navigation, and more. Although Renaissance academies and other learned societies were, on the whole, open to innovation, they were usually not predisposed toward teaching applied knowledge. Manual work was traditionally considered "servile"—despite the classification of seven "mechanical arts" parallel to the seven liberal arts taught in universities: weaving, building, navigation, agriculture, hunting, riding, and medicine. Only the latter was taught at the university, and at a theoretical level. Renaissance applied knowledge, being hands-on practice, is relatively little documented, and because guilds monopolized the trades and their teaching, the juridical aspect is the best documented.

One of the privileges of guilds was to train apprentices and to confer the artisan's license, permitting exercise of the craft, and the master's license, permitting to teach it. The workshop was the nucleus of production and training. The members of a workshop were roughly divided into three classes: masters, apprentices, and boys. The master owned the workshop, was a member of the guild, held the secrets of the trade, and had the right to apply and teach them. He possessed the necessary tools, bought the raw materials, and sold the finished products. He employed apprentices and boys who frequently lived under the same roof.

The apprentices learned to become craftsmen of masters in their own right. The guild's statutes set down the rules, including the number of apprentices a master might teach and how long the apprenticeship was to last. To learn a trade, the apprentice's family signed a contract with a master; sometimes, the master paid the apprentice for relatively menial work. If the apprentice was the son of the master artisan—as it often happened—the rights were passed down through inheritance. The apprenticeship was elementary and hands-on, even though later it could become so sophisticated as to require complex theories, making it a precursor of modern higher technical or technological teaching. The boys, finally, carried out the simpler tasks that did not require any special proficiency.

Learning had to be stressful

Guilds aimed—as modern corporations still aim—at eliminating competition, or at least limiting it; one way to do this was (and still is) to keep non-members from practicing the trade. An indirect method was to limit the number of apprentices, prolong the apprenticeship, and weigh it down with activities—at times irrelevant. In other words, learning could become stressful. At the end of it, if the apprentice was deemed qualified, he was given the right to practice and teach his trade, which included the right to inflict on his new apprentices the same stress he had endured. Still today, universities tend to encumber students' curricula with minutiae, or limit the number of enrolled students (why else impose a *numerus clausus* at medical schools, if not to limit the competition?), and students must—almost by

definition—learn under stress. All these factors, partly a legacy of corporatism, are the source of tremendous ambiguities that continue to burden
modern learning and the university.

Because guilds aimed first and foremost at the preservation of the privileges of their members, they exercised a primarily conservative influence.
Nonetheless, the crafts and trades progressed, paving the way for new fields
requiring trained craftsmen. At first, the emerging generations of craftsmen
had to adapt to the circumstances at hand and find sanctuary among the
existing guilds: in Florence, for example, painters belonged to the guild of
physicians and apothecaries (*Arte dei medici e speziali*) because the latter
also sold colors. They were also frequently members in the Confraternity of
St. Luke (*Compagnia di San Luca;* the evangelist Luke was the patron of the
artists), which, however, was not a guild. Goldsmiths belonged to the guild
of silk weavers and merchants, as did the gold weavers and goldbeaters.
Architects belonged to the guild of stonemasons and woodcarvers, as did
the carpenters and sculptors.

How were these craftsmen trained? Various sources, in particular related
to the fine arts, have provided information on the curricula of Renaissance
apprentices. One of the oldest books, *Il libro dell'arte* (*The Craftsman's
Handbook*) by the Tuscan painter Cennino Cennini (c. 1370–1440), was
extensively used as a manual in painters' workshops and, thus, offers precious testimony. It stipulates long and rigorous training, inter alia:

> Know that there ought not to be less time spent in learning than this:
> to begin as a shopboy studying for one year, to get practice in drawing
> on the little panel; next, to serve in a shop under some master to learn
> how to work at all the branches which pertain to our profession; and to
> stay and begin the working up of colors; and to learn to boil the sizes,
> and grind the gessos; and to get experience in gessoing anconas, and
> modeling and scraping them; gilding and stamping; for the space of a
> good six years. Then to get experience in painting, embellishing with
> mordants, making cloths of gold, getting practice in working on the
> wall, for six more years; drawing all the time, never leaving off, either
> on holidays or on workdays. And in this way your talent, through much
> practice, will develop into real ability. Otherwise, if you follow other
> systems, you need never hope that they will reach any high degree of
> perfection. For there are many who say that they have mastered the
> profession without having served under masters. Do not believe it, for
> I give you the example of this book: even if you study it by day and by
> night, if you do not see some practice under some master you will never
> amount to anything, nor will you ever be able to hold your head up in
> the company of masters.[1]

According to Cennini, an apprentice should last no less than thirteen years;
this is almost certainly an exaggeration, probably dictated by the author's

perfectionism. However, the learning by doing—long or short—began, for artists as for other craftsmen, with elementary manual tasks: apprentices ground the colors, cleaned the paintbrushes, prepared the tables and canvasses; they continued on to more advanced duties, such as reproducing compositions from the cartoons onto paintings, creating the draperies and secondary portions of figures, all the way to executing entire works on the basis of simple sketches or verbal indications. The advantage of this learning was to revolve more around the practical teaching by a master than around a mere textbook.

The advanced apprenticeship, of artists at least, seems to have included the reproduction of models, studies, or works by the master; sometimes apprentices were even entrusted to make entire works on their own, to which the master added the final touches. This practice contradicted the underlying principle of the guild, which protected the originality of the works; it was, thus, a step toward knocking down the barriers, as well as the affirmation of the centrality of the figure of the master as teacher. In other words, it was a step towards more open learning institutions. Reproductions, in Florence at least, were made primarily on Sundays, when there were no obligations imposed by the workshop. At the end of this study period, the apprentice presented a masterpiece—a practice similar to an academic thesis. This was the initiation process to becoming a Master of Art (mechanical, of course): successful candidates became members of the guild, with the right to exercise the craft and train students.

Thus were the great protagonists of the Renaissance trained. Brunelleschi was trained in a goldsmith's workshop. Manetti describes the former's education and avoidance of university studies with the following words:

> Following the general custom of men of standing in Florence, Filippo learned to read and write at an early age and to use the abacus. He also learned some Latin; perhaps because his father, who was a notary, thought of having him follow the same profession, since very few men in that period took up Latin—or were made to take it up—unless they expected to become a doctor, notary, or priest.
>
> (Manetti 1970, 38)

Likewise, Lorenzo Ghiberti (1378–1455) studied in the workshop of goldsmith Bartoluccio Di Michele. Ghiberti is famous for the panel of the Sacrifice of Isaac, now in the museum of the Bargello in Florence; thanks to this masterpiece, Ghiberti won the competition to design doors that would be placed on the north side of the Florence baptistery (Brunelleschi was among the competitors). Ghiberti is also author of *Commentari* (1447), an autobiographical treatise on the theory and technique of the arts.

Mariano di Jacopo, known as "Taccola" ("jackdaw," c.1382–1453), a great Renaissance artist and engineer, attended the workshop of sculptor Jacopo della Quercia (c. 1374–1438). Masaccio (1401–1428) most probably studied in the workshop of his paternal grandfather, Mone di Andreuccio, a carpenter and furniture builder. Piero della Francesca (c. 1412/1417–1492) apprenticed in the workshop of Domenico Veneziano (c. 1410–1461). The list of artists and craftsmen learning in workshops is long.

Besides teaching the craft, the master legitimized his pupils' practice, both present and future, and Renaissance biographies of artists at times extolled the contact between master and pupil, often embellishing it with imaginary details. Similar emphasis can be found, by the way, in early biographies of scientists, which suggests that they were considered on a par with artists (see Segre 1998b).

A Historiographical Caveat: Renaissance biographies are valuable sources but must be considered in their literary context

To be sure, the emphasis on the relation between the apprentice and his master in contemporary biographies may be not only the result of the importance given to the figure of the master, but also dictated by classic literary traditions, such as that of Diogenes Laertius. This historiographic caveat must be taken in due consideration. Ernst Kris (1900–1957) and Otto Kurz (1908–1975) identify recurring themes in these biographies, noting a series of leitmotifs that exalt the childhood of the artist and the effect his work had on the public (Kris & Kurz 1934 and 1979). Kriss was a psychoanalyst and a collaborator of Sygmund Freud (1856–1939), and Kurz was a pupil of art historian Julius von Schlosser (1866–1938). They pay particular attention to the birth of the artist, as though the son of a god had come into the world; they connect the apprentice to a famous master, who was often encountered by pure chance, although the talent of a young genius is innate and the master's role is more that of putative father. In his *Commentari*, for example, Ghiberti tells that Giotto (c. 1267–1337) apprenticed under Cimabue. Although, despite lack of evidence, the two artists may have met in the course of their lives, Giotto, for certain, did not apprentice under Cimabue. Yet Vasari, in his *Lives*, embellishes the story with imaginary details. He also describes the training of Leonardo da Vinci (1452–1519), whose father, Piero, transferred him from the elementary school, where students were taught to do sums with an abacus, to the workshop of Andrea Verrocchio (1436–1488):

> Thus, in arithmetic, during the few months that he studied it, he made so much progress, that, by continually suggesting doubts and difficulties to the master who was teaching him, he would very often bewilder him. . . . Ser Piero, having observed this, and having considered the loftiness of his intellect, one day took some of his drawings and carried them to Andrea del Verrocchio.[2]

Similarly, Vasari describes how Michelangelo's father, Lodovico Buonarroti, sent his son to the workshop of Domenico Ghirlandaio:

> Lodovico, perceiving that he could not divert the boy from giving his attention to design, and that there was no help for it, and wishing to derive some advantage from it and to enable him to learn that art, resolved on the advice of friends to apprentice him with Domenico Ghirlandajo [. . .] as may be seen from an entry by the hand of Lodovico, the father of Michelagnolo, written in one of Domenico's books, which book is now in the possession of his heirs. That entry runs thus: "1488, I record, this first day of April, that I, Lodovico di Leonardo di Buonarrota, placed Michelagnolo my son with Domenico and David di Tommaso di Currado for the three years next to come, on these terms and conditions, that the said Michelagnolo shall remain with the above-named persons for the said period of time, in order to learn to paint and to exercise that vocation; that the said persons shall have command over him; and that the same Domenico and David shall be bound to give him in those three years twenty-four florins of full weight, the first year six florins, the second year eight florins, and the third ten florins; in all, the sum of ninety-six lire."[3]

Vasari, like other contemporary biographers, accounts relatively little additional information concerning the details of artistic training, and biographies are complemented by technical treatises, such as those by Cennini, Brunelleschi, Ghiberti, Piero della Francesca (c. 1416–1492), Leonardo da Vinci, Benvenuto Cellini (1500–1571), Albrecht Dürer (1471–1528), and many others. However, even these treatises do not convey much information concerning the workshop's teaching.

The Renaissance workshop was an incubator of creative learning, and as techniques evolved, vocational schools emerged, where theory was occasionally developed

The workshop was a secular vocational institution that transmitted precious and fundamental knowledge that could hardly be transmitted merely in a lecture or learned from a textbook. As such as it was relatively free from dogmatic thought and had a better chance to enhance a creative process. The drawback was in the lack of the theoretical grounding useful both for problem solving and for developing techniques. The dearth of theory (and of other resources we have today) was compensated through the memorization of techniques, a monotonous activity that did not always promote creativity. Leonardo da Vinci, and many others, deprecated this system of education. Leonardo would have liked to see practice be preceded by science, for "those who love practice without science are like a coxswain entering a ship without rudder or compass."[4] Yet the "learning by doing" of Renaissance workshops, based on direct contact with the master, with the materials, and with reality, is a far cry from lecture-based university teaching, making the workshop a true incubator. Learning became teamwork, which could lessen the tensions often created by subordinating the student to the teacher. Agassi and Ian Jarvie praise this dialectic interaction as an interesting way to improve present-day, modern university teaching.[5] In short, it was an alternative form of teaching, with its own merits and defects. As we shall see, a change in attitude was shortly to come.

As techniques evolved, various levels of vocational schools began to take form in the cities, alongside the workshops and the guilds. One of these, the "abacus school," which sprang up in all the important commercial centers south and north of the Alps, plays a central role in Renaissance learning.

The abacus is a calculating tool that can take many forms. The term *abacus* means sandboard and may be derived from the Hebrew word *Avak*, meaning dust: an early form of abacus was that of a board covered with dust. Despite the rudimental appearance, it can carry out highly complex calculations requesting advanced learning. Even though the primary aim of the abacus schools was to train merchants, they were also attended by students aspiring to other trades. The level of teaching varied, and the curriculum could include subjects that were not strictly related to the vocational teaching, such as grammar. Over time, these schools contributed also to higher knowledge (including that of universities), forming the embryo of modern technological schools. We are acquainted with abacus schools above all thanks to an interesting vein of technical literature inaugurated by Leonardo Fibonacci's *Liber abaci*. As opposed to the *Liber abaci*, written in Latin, these texts were written in a peculiar vernacular and—directly or indirectly—offer information on how the abacus schools functioned (Maccagni 1982).

Abacus schools, operating in cities, were more or less on the level of modern secondary schools. Pupils were enrolled at ten or twelve years of age and attended for several years. They could be of humble origin, or they could be offspring of noble families. Teaching could be either public or private; public lessons, for the less affluent social classes, were paid for by the local municipalities. Private lessons could be offered by teachers who ran their own school, or they could be held for one or more families that employed a teacher. Students learned through direct contact with their teacher and by studying actual examples presented by texts in a colloquial form with questions, answers, and instruction.

The curriculum consisted in learning the Hindu-Arabic numbers, including fractions; the four basic operations; various techniques for making calculations; and the use of the "rule of three" and, naturally, of the abacus. Students used various didactical or practical instruments: they had to memorize many procedures, for example, multiplication tables beyond 10.

Written calculations could be carried out on wooden tablets similar to small blackboards. They learned bookkeeping and how to resolve banking and commercial problems. They learned to calculate compound interest and discounts, profits and losses of capital as a function of time, and other commercial operations, such as barters; to estimate volumes and weights in various units of measure; to calculate money exchanges, percentages of metals in alloys, the paving of wells; to estimate the value of products; to recognize the various commercial markets, and estimate navigational routes and travel times. Occasionally, abacus schools dealt with advanced topics in fields such as geometry or number theory. Thus, although purely formative and practical, they contributed to the development of mathematics.

Abacus teachers did not enjoy the social and financial conditions of university teachers, and perhaps not even that of workshop masters, even though they, too, carried out the dual activities of teacher and freelancer. They could be found in the guilds of the master stonecutters or woodworkers, or in those of other teachers of elementary subjects; sometimes they were able to organize themselves in their own guilds. Moreover, their teaching was less of a sinecure than that university teaching, and contacts with the world outside the classroom were closer, a factor that doubtless encouraged progress. The era of abacus schools is long gone, but a number of their characteristics could still contribute to improving some aspects of the modern university.

From the Renaissance on, techniques, ideas, and people began to cross the barriers of traditional, guild-like institutions in search of more open ones; universities remained, above all, a source of social status

Despite their contributions, most Renaissance craftsmen, engineers, or artists remained relegated to a lower social level than academics, with understandable inferiority complexes. For example, Leonardo da Vinci, who did not know Latin, considered himself, with both pride and regret, an "*omo senza lettere,*" a "man without letters." But even so, the ideas created in workshops influenced prestigious learning institutions such as academies, courts, and even the universities, which, despite their decline, continued, and continue to this day, to project an image of legitimizing authority. Leon Battista Alberti (1404–1472)—the Renaissance genius who contributed to many fields, including architecture, literature, linguistics, philosophy, and cryptography—was instrumental in this process. An ecclesiastic who had studied law at the University of Bologna, he was versatile both in the liberal as in the mechanical ones. In his *De pictura* ("On Painting"), for example, written initially in 1435 in Italian and later in Latin, he aimed to systematize the figurative arts on theoretical (geometrical) basis. His *Ludi matematici,* on the other hand—written between 1450 and 1452—deals with practical physical geometrical problems. As Sergio Rossi (1980, 57–58) remarks, Alberti can be viewed as a pioneer in the process of raising the social status of Renaissance craftsmen.

Some abacus teachers or students, nevertheless, climbed the institutional ladder. Take, for example, Luca Pacioli (c. 1445–1517): his major work, *Summa de arithmetica, geometria, proportioni et proportionalità*, written for didactical purposes and published in 1494 in Venice, was the first general treatise of algebra ever printed. It deals both with pure and applied mathematics. Pacioli is also responsible for perfecting and disseminating the double entry accounting system, which is the basis for bookkeeping in every business today. The style of Pacioli's writings was typical of the abacus literature, indicating that he must have gone through such schooling, although he must have been to a great extent autodidactic as well. He also learned theology to become a Franciscan friar, "thereby acquiring the credentials that allowed him to travel widely and to find academic employment" (Baldasso 2010, 88): he taught in Perugia, Pisa, Bologna—where he probably had Dürer as pupil, later regarded as the greatest artist of the Northern Renaissance—and Rome. He served several courts, and in Milan, he collaborated, among other, with Leonardo Da Vinci.

The most famous abacus teacher and one of the greatest Renaissance mathematicians was Niccolò Fontana Tartaglia (c. 1499–1557). He had a turbulent childhood: he was orphaned at six years of age, when his father, a mail rider, was killed. As though that were not enough, when he was about twelve, during the Sack of Brescia (1512), a soldier sliced with his sword Niccolò's jaw and palate, and the latter stuttered for the rest of his life—hence his nickname, "Tartaglia" ("stammerer"). He did not attend a university; as he relates, as a child he began learning to read and write with the help of a teacher, but when they got to the letter "K," the money ran out, and he had to learn the rest of the alphabet on his own, "in the sole company of an offspring of poverty, called industriousness."[6] As abacus teacher he taught mathematics privately in Verona and in Venice, where he remained almost uninterruptedly for the rest of his life, and he contributed in various fields, including mathematics and ballistics. Tartaglia's most famous discovery is the solution to the cubic equation for which he had a bitter controversy over priority with Girolamo Cardano (1501–1576), a well-known mathematician, physician, astrologer, sorcerer, and academic who taught at the university of Pavia. Cardano did not lower himself to discussion with an abacus teacher: he passed the matter to his assistant, Lodovico Ferrari (1522–1565), who was more or less socially equal to Tartaglia. Cardano himself was imprisoned by the Inquisition in 1570 and had to abandon university teaching.

Abacus teachers furthered the progress of mathematics, despite the difficulties applied mathematicians had in asserting themselves. Over time, an osmosis of ideas and people in search of recognition and prestige between various institutions generated hybrid structures—part workshop and part university. These were the first steps toward technical higher education.

Hybrid higher learning institutions such as the Accademia del Disegno gave rise to technical higher education, a first step towards technological higher learning

Vasari reports (or invents, because, as Camesasca 1966, 240, points out, we lack documentation) how at the end of the 1400s, Lorenzo the Magnificent sponsored an institution to train highly gifted artists, that is, the elite of Renaissance artists. Despite its length, his description is interesting:

> Lorenzo the Magnificent, then, always favored men of genius, and par-
> ticularly such of the nobles as showed an inclination for these our arts;
> wherefore it is no marvel that from that school there should have issued
> some who have amazed the world. And what is more, he not only gave
> the means to buy food and clothing to those who, being poor, would
> otherwise not have been able to pursue the studies of design, but also
> bestowed extraordinary gifts on any one among them who had acquit-
> ted himself in some work better than the others; so that the young stu-
> dents of our arts, competing thus with each other, thereby became very
> excellent, as I will relate. The guardian and master of these young men,
> at that time, was the Florentine sculptor Bertoldo, an old and practiced
> craftsman, who had once been a disciple of Donato. He taught them,
> and likewise had charge of the works in the garden, and of many draw-
> ings, cartoons, and models by the hand of Donato, Pippo,* [* Filippo
> Brunelleschi.] Masaccio, Paolo Uccello, Fra Giovanni, Fra Filippo, and
> other masters, both native and foreign. It is a sure fact that these arts
> can only be acquired by a long course of study in drawing and diligently
> imitating works of excellence; ... Among those who studied the arts
> of design in that garden, the following all became very excellent mas-
> ters; Michelagnolo, the son of Lodovico Buonarroti; Giovan Francesco
> Rustici; Torrigiano Torrigiani; Francesco Granacci; Niccolo, the son of
> Jacopo* Soggi [* The name given in the text is Domenico.]; Lorenzo
> di Credi, and Giuliano Bugiardini; and, among the foreigners, Baccio
> da Montelupo, Andrea Contucci of Monte Sansovino, and others, of
> whom mention will be made in the proper places.[7]

It might be that Vasari dreamed of "elevating" artistic training to an academic level, at least in his writings. Be that as it may, this dream slowly became reality, and mechanics and applied mathematics began to gain ground, also on a social level. Even Martin Luther (1483–1546) urged Germany's burgomasters to introduce mathematics, despite its scant social dignity, into scholastic programs, affirming that he had little to do with mathematics: "If I had children and the means, not only would I make them learn languages and history, but also singing, music, and all of mathematics."[8] A relationship that was occasionally problematic, and occasionally synergetic, was created between old and new higher learning and between the individuals and institutions. Subjects initially taught in abacus schools became part of university curricula; abacus teachers and students occasionally found a position in a university or an equally prestigious institution. This process was often welcomed by the princes who endeavored to lessen the power of various traditional learning structures and, at the same time, exploit technical innovation. Workshops made increasingly important contributions to scientific knowledge, constantly striving for social and academic recognition. Nonetheless, the journey was still long and hard. One important step forward was the foundation of the Florentine *Accademia del Disegno*.

As early as 1540, Duke Cosimo I de' Medici named the sculptor Bartolomeo (Baccio) Bandinelli (1488–1560) superintendent of the Works of the Florence Cathedral. Zygmunt Waźbiński (1933–2009) explains the importance of this decision as the first attempt to reorganize Florentine craftsmanship (Waźbiński 1987, 1, 66). The aim was both to assume control and to invest in professional training. "Cosimo had an interest in bringing artists under the same kind of controls that he had already established for *letterati* with the incorporation of the Accademia Fiorentina" (Dempsey 1980, 553). Bandinelli directed a small school for artists that he insisted on calling an academy, probably with the intent of elevating the mechanical arts to the more prestigious level of the liberal arts and himself from the level of a middle-class artisan to that of an academic artist. The initiative led to the creation of similar workshops tied to the court.

In 1563, a further step was made on the institutional level with the founding of the *Accademia del Disegno*, incorporating the Florentine Confraternity of St. Luke; the *Accademia del Disegno* aimed initially to the construction of one of the Medici chapels (the *Sagrestia Nuova*—new sacresty) in the Basilica of San Lorenzo in Florence. Work on this chapel had begun much earlier, in 1520, under the direction of Michelangelo (1475–1564), but was interrupted in 1534 when the latter went to Rome to concentrate on painting the Last Judgement in the Sistine Chapel. The San Lorenzo chapel, however, became a meeting point for artists. In 1555, the chapel's construction was entrusted to Vasari, who, in a letter he wrote that same year to Michelangelo, suggested that his disciples should complete the work: "to make an honorable memorial to the benefit of the public and of Your Excellency . . . many illustrious talents would be employed [and] would give renown to the Academia in your School and in the house of Your Illustrious Excellency, and where each has learned the art."[9] The worksite inaugurated by Michelangelo was transformed into a higher school of art, chartered and with a university frame (in the sense of guild) by Grand Duke Cosimo I.

The 1563 statutes of the *Accademia del Disegno* stated, among other, that the curriculum should include basic mathematical texts: "Euclid, Vitruvius and other mathematicians" (Adorno & Zangheri 1998, 3–26). The teaching should be personalized. Four times a year, artists could attempt to become members of the guild by presenting a work respecting the university tenets. The statutes also speak of the Academy in terms of a *"studio"* (Adorno & Zangheri 1998, 6), and Charles Dempsey (1980, 554) remarks, "Not only is the Academy named a *Studio*, but also it is declared that its membership shall be open to persons from everywhere (thus qualifying it for the title of *studium generale*)."

The founding of a new "university" (always in the sense of a guild) was not entirely problem-free. As previously mentioned, in the beginning the Academy its members belonged to, and depended on a number of guilds: painters, for instance, belonged to the guild of physicians and apothecaries; architects belonged to the blacksmiths' guild. It was not until 1571 that they came to belong to their own "university" of fine art through a specific intervention on the part of Grand Duke Cosimo I.

The *Accademia del Disegno* taught a broad range of subjects: anatomy, at Florence's hospitals; geometry; mechanics; the construction of roads, canals, and bridges; architecture; perspective; music; counterpoint; oratory; and what we would call today "applied chemistry." Besides training professional artists, the academy was a center of innovation, combining theory and practice. To be sure, the corporative interests of the academy soon prevailed its didactical ones, bringing the academy's quick decline (Rossi 1980, 180). Yet it remains an interesting attempt to grant an academic status to the "mechanical arts."

Among the mathematicians associated with it in the 16th century one finds the Dominican Ignazio Danti (1536–1590), astronomer and cosmographer, in charge of the grandiose, never accomplished, plan to bore a canal connecting Florence to the Adriatic and placing the city in the middle of a communication waterway from one side of the Italian peninsula to the other. Danti later taught at the University of Bologna.

One other prominent mathematician active within the *Accademia del Disegno* was Ostilio Ricci (1540–1603), Galileo's private teacher. A patrician and member of the Tuscan court, Ricci, too, did not study at a university and may have been a pupil of Tartaglia in Padua. In 1586, under the Grand Duke Francesco de' Medici (r. 1574–1587), son of Cosimo I, he became court mathematician and tutored the pages. His private lessons sparked such great passion for mathematics in Galileo. Direct contact with the new and fascinating aspects of mathematics seemed to be more interesting than the liberal art lessons at the Pisa University, whose topics were established by a set curriculum. In this manner—and not through university teaching—the seeds of modern science were sown.

One interesting and successful 16th-century engineer related indirectly to Galileo was Giovanni Battista Benedetti (1530–1586), a Venetian patrician and pupil of Tartaglia. Although he did not pursue university studies, he became a court mathematician, first under Ottavio Farnese, Duke of Parma and Piacenza (r. 1547–1586), and then under Emmanuel Philibert, Duke of Savoy (r. 1553–1580). His mechanical studies produced results that are curiously similar to those produced a few decades later by Galileo. One interesting result was his refutation, through a thought experiment, of Aristotle's affirmations in *On the Heavens* (*De Caelo*) and Physics (*Physica*), that the weight of a body determines the velocity of its fall (Cooper 1935, 59–68). In an early work, Galileo describes a very similar thought experiment, even if there is no indication that Galileo was familiar with Benedetti's works. The reasoning is simple and ingenious: if a body falls at a velocity that is proportional to its weight, two bodies that are united but of different weights should fall at a velocity greater than that which each would achieve individually. According to this same Aristotelian reasoning, the smaller body should, nonetheless, slow down the larger body, and this leads to a contradiction. Therefore, the Aristotelian law, according to which the falling velocity of an object is proportional to its weight, is logically unacceptable. The refutation is provided both within technical tradition familiar to Benedetti and within university tradition to which young Galileo adhered.

Benedetti, as an engineer, remained outside the contemporary university establishment, but Galileo learned from various traditions and combined them to perfection. We find him in universities, in academies, in workshops, and at court. His ideas not only synthesized the knowledge coming in from all these directions, he also had the courage to break down traditional, social, and disciplinary barriers, a risky undertaking but scientifically fruitful. Indeed, he ended his days under the custody of the Inquisition, while others ended up burning at the stake. Galileo's last scientific work, *Discourses and Mathematical Demonstrations Relating to Two New Sciences* (1638), begins by mentioning the arsenal in Venice:

> The constant activity which you Venetians display in your famous arsenal suggests to the studious mind a large field for investigation, especially that part of the work which involves mechanics; for in this department all types of instruments and machines are constantly being constructed by many artisans, among whom there must be some who, partly by inherited experience and partly by their own observations, have become highly expert and clever in explanation.[10]

Galileo pays tribute to technical science: the two new sciences are the "pure" science of motion and the mechanical science of the resistance of bodies. But here we enter full force into the scientific revolution of the 17th century, to which the next chapter is dedicated. Let me here only mention that Evangelista Torricelli (1608–1647), Galileo's famous assistant and one of the leading mathematicians of his century, too, was associated with the *Accademia del Disegno*.

Academies similar to the *Accademia del Disegno* were soon founded, such as the homonymous academy founded in Perugia in 1573. The recurring challenge was to free themselves from the domination of the guilds and to obtain more prestigious social recognition. In 1608, the *Accademia Delia* was founded in Padua, introducing local nobility to military and civilian engineering, which, "although by rank and dignity were separate from the University, nonetheless held to the same aims and were animated by the same spirit" (Favaro 1966, 2, 237). The Renaissance technical schools were influential because they were the prelude to the more advanced schools founded in the 1700s, above all in France, and which, in turn, were the precursors of the modern polytechnic institutes and the modern university (see Chapter 11).

CONCLUDING REMARK

Despite their low social standing, Renaissance workshops made substantial contributions not only to the progress of science, but to that of teaching, as well.

BIBLIOGRAPHICAL NOTES

Zygmunt Ważbiński's *L'Accademia Medicea del Disegno a Firenze nel Cinquecento* (1987) is a detailed work in two volumes, including documents, concerning the dawn of the *Accademia del Disegno*. Sergio Rossi's *Dalle botteghe alle accademie* (1980) is a detailed study attempting to describe developments in the social status of artisans within a Marxist frame of class struggles. The result was elevating the status of only part of them, that is, the more "intellectual" artists. Charles Dempsey's "Some Observations on the Education of Artists in Florence and Bologna During the Later Sixteenth Century" (1980) is a useful outline concerning the same Academy in English. Lane Cooper's *Aristotle, Galileo, and the Tower of Pisa* (1935) is a deliciously amusing book claiming, on the basis of collected and quoted sources, that Galileo might not have performed the Leaning Tower Experiment exactly as according to enduring legend.

NOTES

1. Chapter CIIII, quoted from http://www.noteaccess.com/Texts/Cennini/5.htm, accessed June 11, 2012.
2. Vasari, *Le Vite*, part III, http://members.efn.org/~acd/vite/VasariLeo.html, accessed June 11, 2012.
3. Vasari, *Le Vite*, part III, http://members.efn.org/~acd/vite/VasariMichelangelo1.html, accessed June 11, 2012.

4. Pevsner 1973, 35: "Quelli che s'inamorano di pratica senza scientia sono come li nocchieri che entran in naviglio senza timone o bussola."
5. Agassi and Jarvie (1987), "The Rationality of Dogmatism," pp. 431–445. See pp. 442–443. Agassi's intellectual autobiography (1993) is subtitled *In Karl Popper's Workshop*, an indication of interactive teaching in Popper's lecture halls and of the possibility of applying artisan workshop-type teaching methods even in the field of philosophy. But as the book itself suggests, Popper theorized workshop methods more than he actually practiced them, and the learning process under the grand maestro was not immune to tension.
6. Tartaglia 1554, 139: ". . . *ma solamente in compagnia di una figlia di povertà, chiamata Industria.*" http://echo.mpiwg-berlin.mpg.de/ECHOdocuView?view Mode=text&viewLayer=[%27dict%27%2C+%27search%27]&url=/permanent/ archimedes_repository/large/tarta_quesi_042_la_1554/index.meta&query=in dustria&pn=1&queryType=fulltextMorph, accessed October 20, 2013.
7. Vasari, *Le Vite*, http://members.efn.org/~acd/vite/VasariTorrigiano.html, accessed June 12, 2012.
8. See Olschki 1965a [1919], pp. 420–421 and note 3: "Wenn ich Kinder hätte und vermöcht's, sie müssten mir nicht allein die Sprachen und Historien hören, sondern auch singen und die Musica mit der ganzen Mathematica lernen."
9. Waźbiński 1983, 62, note 29; Waźbiński 1987, 1, 86, note 52: "cercando di fare «qualche onorata memoria al beneficio pubblico e di V.E.» [. . .] «esercitarebbe tanti illustri ingegni [et] darebbe la fama alla Accademia nela sua Scuola et nella propria casa di V.E.I., et dove ciascuno ha imparato l'arte [. . .]»"
10. Galileo Galilei, *Discourses and Mathematical Demonstrations Relating to Two New Sciences* (1638), http://galileoandeinstein.physics.virginia.edu/tns_draft/tns_001to061.html, accessed June 15, 2012.

9   The Advent of Science

The scientific revolution took place outside the university

The 16th century was a golden age for learned societies and, not by chance, the beginning of the scientific revolution. In 1543, Nicolaus Copernicus (1473–1543) published his *De Revolutionibus* ("On the Revolutions"), inaugurating a new world view: against intuition and, for some, in contradiction to the Holy Scripture, he proposed that the Earth orbits around the Sun. This masterpiece, written in Latin, is a complex mathematical treatise.

Copernicus was not an academic, even though he had attended various universities. He was, however, a canon (a member of the clergy not ordained as a priest), with administrative duties at the chapter of Warmia, in northeastern Poland. He had probably studied astronomy as part of the basic liberal art curriculum at the University of Krakow, after which he went to study canon law in Bologna and Padua. In Padua he also studied medicine. He graduated, always in canon law, in Ferrara, probably because graduating there was less expensive. He even got a sinecure as *scholaster* (dean) at the Collegiate Church of the Holy Cross in Breslau, but he never taught there. Instead, he practiced medicine and dabbled in astronomy. In any case, he would have had a hard time teaching his bizarre theory at a university.

Many people, at that time, opposed the Copernican theory, including another amateur astronomer-astrologer, the wealthy Danish aristocrat Tycho Brahe (1546–1601). Like Copernicus, Tycho had studied at a number of universities: Copenhagen, Leipzig, Wittenberg, Rostock, Basel, and Augsburg. Tycho was not a cleric and became famous as an astronomer despite the universities, and no thanks to them.

Tycho's father would have liked his son to undertake a political career, but Tycho became infatuated with astronomy after having observed a partial solar eclipse. His tutor, Anders Sørenson Vedel (1542–1616), a cleric later to become a well-known historian, was given the task of squiring him around the universities of Europe in the hope of bringing the young man to reason—all in vain. Obviously, this story smacks of the legendary, as do many contemporary biographies of scientists or artists. What is certain is that Tycho Brahe belonged to that class including noblemen or wealthy commoners who considered astrology and science respectable leisure occupations. When in 1574 Tycho was asked to lecture on astronomy at the University of Copenhagen, the difficulty that soon emerged was that he belonged to the wrong caste: it "would not have been appropriate for Tycho as a nobleman to lecture at a university for this would have been seen as intruding into the privileges of another estate" (Thoren 1990, 79).

Tycho, nevertheless, was much more than an amateur; he was an enthusiastic observer of the heavens, with a mystical vocation and an extraordinary ability to build precision instruments. He gained the support of King Frederick II of Denmark (r. 1559–1588), who in 1576 granted him the island of Hven (not far from Copenhagen, today in Sweden) as his estate, where he built an out-and out research center. It consisted of two astronomical observatories: the first, "Uraniborg" ("Castle of Urania," the muse of Astronomy), whose architecture reflected a Neo-Platonic and hermetic view of the universe, had observation rooms equipped with highly precise astronomical instruments, and rooms for alchemic research. The site comprised a library, a printing press, guest quarters, and every amenity for its residents, including a luxurious residence for Tycho himself. The second observatory, "Stjerneborg" (Castle of the stars), had large astronomical instruments located underground and protected from the wind by special cupolas. Tycho's celestial observations were the highpoint of pre-telescopic astronomy, even if he kept many results secret, in line with the mystical tradition to which he belonged. Be that as it may, Tycho Brahe was a great scientist.

One of Tycho's important contributions to science that he did not keep secret was his proposal of a cosmological system, attempting to solve the physical and Scriptural objections to the Copernican system: the Sun and the Moon orbit around the Earth, while the planets orbit around the Sun. Not only was this a compromise between two cosmologies, but, more in general, between tradition and innovation, a clear (albeit camouflaged) break with Aristotelian cosmological tradition. In the Aristotelian universe, the planets were moved by a complex system of crystalline, concentric spheres—both physical and metaphysical—around the earth. Tycho Brahe, by letting the planets orbit around the sun, ultimately trashed this very system.

Tycho's observatory was a center of astronomical research, hosting young researchers, who lived there as though at a college. Roughly thirty scholars, from various countries and universities, came to work or learn at Hven. Tycho paid little attention to their university credentials, as testified by a sarcastic note in a letter dealing with a candidate to work with him: "Whether or not he has obtained the M.A. degree, is immaterial to me. I would prefer that he really be a master of arts, rather than just have the degree" (Thoren 1990, 197)—an acute observation that indicates the gap, then as now, between university studies and fieldwork. Victor Thoren, in his detailed biography of Tycho Brahe, comments that "Most of the personnel problems that Tycho experienced [. . .] were due to the essential incommensurability of the university education of the day and the research Tycho was doing on Hven" (Thoren 1990, 197).

We do not know how exactly learning was produced or shared in Tycho's singular workshop. Hot-tempered Tycho may have acted as a teacher-master, and working under him must have been all but easy, although enthralling. His most talented collaborator was Elias Olsen Morsing (c. 1550–1590), a Dane who lived at Hven from 1583 until his death in 1590. Another key collaborator was the goldsmith Hans Crol, who died the following year (1591). The two were replaced by Christian Sørensen (1562–1647), also known as "Longomontanus" (a Latinized form of the name of the village of Lomborg in Denmark, where he was born), the only one out of Tycho's students to obtain a university position as an astronomy professor. There was also the enterprising Dutchman Willem Janszoon Blaeu (1571–1638). After completing their period in Tycho's workshop, the apprentices could aspire to respectable non-academic positions.

The "kingdom of the heavens" at Hven, combining technical expertise, research, and scientific training, did not last long. In 1597, after the death of Frederick II, Tycho had to abandon the island and wandered across Europe with assistants, servants, instruments, books, and equipment in a caravan that Arthur Koestler (1905–1983) labels a "private circus" (Koestler 1959, 295). He ended up under the patronage of Emperor Rudolph II of Hapsburg (r. 1575–1612) at the castle of Benátky, near Prague, then capital of the Holy Roman Empire. One of his assistants there was young Johannes Kepler (1571–1630), one of the greatest astronomers of all time and a pillar of the 16th-century scientific revolution.

Like Copernicus and Tycho, Kepler, too, had studied at a university. More precisely, he studied theology at Tübingen. He taught astronomy in Graz before becoming the mathematician at the ducal court in Styria. After assisting Brahe in Prague, he succeeded him as imperial mathematician. He set the theoretical foundations of modern astronomy as a courtier, not an academic. Kepler delved into the study of Brahe's highly precise observations of the orbit of Mars and deduced that the planet's orbit was elliptical and not circular, in opposition to everyone else, including Copernicus and his contemporary, Galileo. Actually, all the planets move in ellipses with the Sun at one of the two foci. To be sure, their orbit is not exactly elliptical due to the mutual gravitational attraction, discovered only later by Newton. Kepler published this finding in his *Astronomia Nova* (1609). In the introduction to this work Kepler rejects the arguments against heliocentrims and the motion of the earth on scriptural basis:

> The holy scriptures, too, when treating common things (concerning which it is not their purpose to instruct humanity), speak with humans in the human manner, in order to be understood by them. They make use of what is generally acknowledged, in order to weave in other things

more lofty and divine. [. . .] While in theology it is authority that carries the most weight, in philosophy it is reason.

(Kepler 2008, 19, 25)

Similar arguments, separating between religion and science, were to be raised soon, eloquently, by Galileo, another scientist who distanced himself from the universities as soon as he could.

Although Tycho's observatory on the island of Hven was the only one of its kind at the time, the number of learned societies dedicated to science increased, as we will see more explicitly in the next chapter. In 1603, the young and wealthy prince Federico Cesi (1585–1630) founded and sponsored in Rome the *Accademia dei Lincei* (after the lynx, whose sharp vision symbolizes the observational ability of scientists; the name may have been inspired by Della Porta's *Magia*, see Clubb 1965, 37). It had only four founding members, and their activities consisted in teaching each other scientific matters. In 1610, they were joined by the elderly Della Porta himself, but it was Galileo, who became a member in 1611, who brought fame to the small academy, which, in turn published his works and gave him its unwavering and unconditional support. The academy ceased its activity after Cesi's death and shortly before Galileo's trail in 1633. Yet later, in virtue of its illustrious member, various attempts were made to resuscitate and officialize the academy; in 1875, one of these attempts led to the founding of the royal, national academy of Italy. Today, the *Accademia dei Lincei*, although no longer royal, gathers the elite of Italian intelligentsia; it retains only the name and the luster of its 17th-century progenitor. And in parallel, in 1936, under Pope Pius XI (r. 1922–1939), the Pontifical Academy of Sciences was established, having the Vatican as seat and gathering many outstanding scientists from all over the world. With no intention of minimizing the noble ends of these present-day academies, like many other contemporary national academies, they primarily confer prestige upon their members, their humble ancestors notwithstanding. Most academies, as will be seen in the next chapters, ended up taking some kind of guild-like form, with all the advantages and drawbacks. The intensification of the social and political role of the academies brings us back to the central problem: the tension between the intellectual function and the social function of a scientific institution. This tension is also well reflected in the life and the science of one of the founders of modern science: Galileo Galilei.

Galileo Galilei was one of the first modern scientists: he drew his knowledge from different intellectual traditions

. Galileo Galilei (1564–1642) was the son and pupil of Vincenzo Galilei (1520–1591), a lutenist, composer, and music theorist of vast erudition who was, among other, interested in the revival of ancient Greek music. Thus, already at a young age, Galileo was exposed to various contemporary intellectual currents, in particular regarding the arts. Like Copernicus, Tycho, and Kepler, he attended the university—at Pisa—as a liberal arts student (*OG* 19, 32). His father, according to his biographers Vincenzio Viviani (1622–1703) and Niccolò Gherardini (d. 1678), would have preferred to see him become a physician, but Galileo abandoned the university before receiving a degree, preferring to take private lessons from Ostilio Ricci, mentioned in the preceding chapter (*OG* 19, 602, and 635). Ricci taught mathematics (Euclid, Archimedes) and its applications as well as the theory and practice of perspective. His textbooks included the writings of Tartaglia and Alberti's *Ludi matematici*. Thomas Settle (1971) suggests that Ricci most likely introduced Galileo to the works of Alberti. Besides Galileo, Ricci's pupils included the painter Ludovico Cigoli (1559–1613) and Don Giovanni de' Medici (1567–1621), the natural son of the Grand Duke Cosimo I and a future military engineer. Galileo's education thus reflected at least three distinct intellectual traditions: musical (i.e., the artistic tradition), academic (i.e., the university tradition), and technical (i.e., workshops), the latter winning out over the others.

Galileo did not hold the university in high esteem, and if one looks at the list of lecturers at the university of Pisa for the 1584–85 academic year (*OG* 19, 32–35), one finds that most of them were clerics.[1] Around 1590, when Galileo was 26 years old, he composed a poem in triplets titled "*Contro il portar la toga*" ("Against wearing the academic dress"; *OG* 9, 213–223), which waxed critical of contemporary academics. Galileo did not criticize academics only for their lifestyle, but also for their methods of investigation:

> I am really caused great trouble by those people
> Who set about examining the Supreme Good
> And until now have not come to grips with it.
>
> And I project with my imagination
> That this only comes about
> Because it is not where they are looking for it.
>
> These learned men have never understood properly,
> Never ventured on the right track
> That might lead them to the Supreme Good.
>
> For the reason that, in my opinion,
> Whoever wants to understand the truth of something
> Has to use his fantasy,
>
> Play with ideas, and find the key to the puzzle;
> And if you cannot reach it by the most direct route,
> There are a thousand other ways to help you.
>
> (Reynolds 2002, 55)

Not only does Galileo emphasize the importance of using one's imagination, he also puts forward a specific methodology of science:

> Ways of rational investigation vary greatly;
> But to find Good I have experienced
> That you have to proceed via its converse:
>
> Look for Evil, and there it is, plain as day;
> For Supreme Good and Supreme Evil
> Are as alike as chickens at the market.
> (Reynolds 2002, 55)

Galileo appeals for broad-ranging study based on trial and error with imagination as essential ingredients in science, a method that was later endorsed by scientists and philosophers such as David Brewster (1781–1868) and William Whewell (1794–1866), and, four centuries later, by Popper, who was the more extremist as he denied certitude.

Despite his scorn, Galileo taught at the university for many years: first in Pisa (1589–1592), then, for eighteen years, in Padua (1592–1610). Padua had been under Venice's jurisdiction since 1405, and the Venetian authorities did their best to ensure its professors and students the greatest possible freedom of thought, turning the city into a cosmopolitan center. The many students included Jews and Protestants from all over Europe, and this was unusual because universities were generally denominational until much later. The city's proximity to Venice offered many advantages: overseas contacts, a lively naval dockyard, sources of innovative ideas, and, last but not least, a prolific publishing industry. The university boasted an Aristotelian tradition in philosophy and was still on the cutting edge in medical innovation. The medical curriculum included both theory and practice as well as a two-year course in philosophy offering many debates. Padua promoted mathematical branches at that time still generally neglected, such as optics, mechanics, and cosmography (i.e., astronomy and geography): these were to develop into fields of modern science.

Not all the Paduan students frequented the university for study purely in and of itself. As in universities generally, then as now, many were only interested in the prestige it conferred. As described by a contemporary,

> Not all who call themselves students and go to Padua, go there to study letters. At most, the majority of the French students learn to ride, to dance, to practice the use of all sort of weapons, and Music, and finally they learn Italian customs and courtesy, of which they are enamored, and other similar virtues, rather than for reasons of letters. Thus, this is why they chose to study in Padua, which is abundant with professors highly excellent in every sort of magnificent and illustrious virtue.
> (Favaro 1966, 1,142)[2]

Those "highly excellent professors" also gave private lessons, potentially offering the advantages of a workshop atmosphere with freedom to teach innovations. Private lessons differ from tutorial because they are utterly independent of university tuition. The latter, as seen in chapter 6, were introduced in as early as the 14th century in Oxford. Private lessons obviously began when some aristocrats could hire private tutors full time and others could do so only on a limited basis. They also gave professors the chance to round out their salaries, a legitimate practice, if kept within the limits of decency. This decency, however, progressively waned. As is also the case in almost every university today, to obtain their degree, Paduan students had to accumulate credits (called *ponti* or "points"). Modern universities not only persist with this deplorable practice, but even increase it, disregarding that, as argued by Popper in *Objective Knowledge* as well as by other philosophers, knowledge cannot be quantified, much less measured. In Padua, these *ponti* were already a lure for corruption: they were increasingly awarded outside the lecture hall, presumably in exchange for money. This led the university authorities to decree in 1614,

> No lecturer [*Dottor Leggente*] may read credits [*ponti*] in private to any student, under those penalties, etc.: and in spite of such a good regulation, a number of lecturers or students began making a private school [*Accademia*] of said credits, reciting [*recitandone*] and exercising at the residence of lecturers, and of others who are not lecturers, without attending university lectures in current subjects, so that they graduate in a very short time without having any knowledge whatsoever.
>
> (Favaro 1966, 1,140, my translation)

The expression "reciting" indicates the extent to which the material was still learned by heart, as common in an oral culture as well as still in some modern universities. More generally, we have a basic problem related with university studies yesterday and today. If a student studies in order to grow intellectually or professionally, lessons or workshops are almost always interesting and instructive. If learning is exploited for social purposes and students study only to pass exams or accumulate credits, marks, or degrees alone, then interest diminishes and study can degenerate into stress, alienating the student's interest in the subject being studied and indirectly hampering science. The problem is still current, as noted by Popper:

> Instead of encouraging the student to devote himself to his studies for the sake of studying, instead of encouraging in him a real love for his subject and for inquiry, he is encouraged to study for the sake of his personal career; he is led to acquire only such knowledge as is serviceable in getting him over the hurdles which he must clear for the sake of his advancement. In other words, even in the field of science, our

methods of selection are based upon an appeal to personal ambition of a somewhat crude form.

(*OS* 1, 135)

Much can still be done to make study more genuine, honest, and pleasurable. However, the relationship between science and society remains basically problematical. Science cannot progress without the help of institutions such as the university, but the institutions can get corrupted; only judiciousness and intellectual honesty and foresight can avoid this problem for the moment. To return to Galileo, in his time, Padua was a prestigious university in decline.

In Padua, Galileo, although unmarried, had a full family life with his mistress, Marina Gamba (c. 1570–1612), who bore him two daughters and a son. Why did Galileo not marry? Most probably because university professors belonged to the clergy and did not marry. Galileo, a layman, was an exception, but marrying could have harmed his reputation.

As to teaching, Galileo's public lessons, which dealt above all with mathematical topics such as Euclidean geometry or Ptolemaic astronomy, attracted many students. Yet a great deal of his scientific activity, including private teaching, took place at home, where he hosted students, gave lessons, carried out experiments, and built instruments. His geometric and military compass, a device for making measurements and calculations, required both artisan craftsmanship and mathematical knowledge. Galileo manufactured them in a great number. He also constructed his telescope in his home laboratory, not an easy undertaking because the mathematical theory of optics had yet to be developed. As for his private lessons, Galileo kept a register listing the names of his many pupils, including foreigners, and the subjects he taught. They included the following: the use of the geometric and military compass; the theory of fortifications; Euclidean geometry; perspective; the theory of sphere, which was closely linked to cosmography; arithmetic; and geodesy and mechanics, a science in evolution, part philosophy and part engineering, whose mathematical and modern foundations were laid by Galileo himself (Favaro 1966, 1, 184). The private curriculum went well beyond the university's, even though, as Antonio Favaro (1847–1922), the modern editor of Galileo's works, makes note, Galileo did not yet dare teach risky subjects, such as Copernicus's theory. He indicated this in his letter to Kepler of August 4, 1597:

I have written many arguments in support of him [Copernicus] and in refutation of the opposite view—which, however, so far I have not dared to bring into the public light, frightened by the fate of Copernicus himself, our teacher, who, though he acquired immortal fame with some, is yet to an infinite multitude of others (for such is the number of fools) an object of ridicule and derision.

(Koestler 1959, 336)

Starting in 1609, and thanks to his telescope, Galileo made his first sensational astronomical discoveries: the satellites of Jupiter, the irregular surface of the Moon, the starry composition of the Milky Way, and the phases of Venus. These discoveries made him so famous that he was able to abandon university teaching and become a courtier. In a letter he wrote in 1610 to Belisario Vinta, Tuscan Secretary of State, he describes his university activities in Padua and the reasons he would like to abandon them and join the Tuscan Court:

> My duties do not occupy more than 60 half hours per year, and not in such a stringent manner that, for any impediment of mine I may not, and with no disadvantage, also interpose many empty days. The rest of the time I am completely free, and absolutely *miei iuris*. But since both private lessons and domestic students would be an impediment to me and a hindrance to my studies, I would like to be completely free of the one and to a great degree relieved of the other. But, should I decide to return to my homeland, I would wish that the first intention of His Highness be to give me the leisure and comfort to complete my works, without having to occupy myself in reading.
>
> Nor would I wish that this cause His Highness to believe that my efforts to be less profitable to the scholars of the profession, on the contrary they will absolutely be more so; because in public lessons one cannot read other than the basic elements, to which many are suited, and this reading is only an impediment and of no help in leading to the completion of my works, which I believe will not be the least among things pertaining to the profession.
>
> (*OG* 10, 350–351, my translation.)[3]

The last paragraph points out to the (questioned) habit, yesterday as today, to lecture only the accepted basics. (Einstein, as seen in chapter 1, was an exception).

Although Galileo's university duties required only a modicum of teaching, they remained a hindrance to his scientific work. Moreover, in order to overcome the traditional social barriers hampering his creative activity, Galileo officially demanded the title of "philosopher" besides that of mathematician, granting him higher socio-intellectual status. Philosophers had the right to deal with the essences, a right that was generally denied to mathematicians.

Galileo left Padua, and, from that moment on, his scientific work took place outside the university. He traveled to Rome to get the blessing of the Jesuit mathematicians of the Roman college, an indication of the prestige that this institution still enjoyed and, more in general, of the ecclesiastical control over science. Despite some scientific disagreement, the friendly welcome reflects intellectual openness. During this visit, however, Galileo joined Cesi's Lincean Academy, which did not normally enroll university people, and for the rest of his career emphasized his being a "Lincean" rather than a university professor.

Unfortunately, the honeymoon was short-lived. Galileo embarked in a campaign in favor of Copernicanism, raising much opposition: the most dangerous criticism came from theologians who argued that the Copernican theory contradicts the Scripture. Galileo replied with his impressive *Letter to the Grand Duchess Christina* (1615), concerning the relations between religion and science.

Galileo was the first to thoroughly discuss, and clearly affirm, the independence of science from faith

The *Letter to the Grand Duchess Christina* displaying, inter alia, Galileo's theological learning concurs that "A multitude of truths contribute to inquiry and to the growth and strength of disciplines," but "The Holy Scripture and nature derive equally from the Godhead, the former as the dictation of the Holy Spirit and the latter as the most obedient executrix of God's orders." Yet, "It is the opinion of the holiest and most learned Fathers that the writers of Holy Scripture not only did not pretend to teach us about the structure and motions of the heavens and the stars, and their shape, size, and distance, but that they deliberately refrained from doing so, even though they knew all these things very well." So, "To prohibit the entire science would be no different than to reject hundreds of statements from the Holy Writ" (Finocchiaro 1989, 87, 93, 94, 103, respectively). Galileo's arguments remain valid today, and Pope John Paul II (r. 1978–2005) labeled the *Letter* "like a short treatise on biblical hermeneutics," adding, "Galileo, a sincere believer, showed himself to be more perceptive in this regard than the theologians who opposed him" (John Paul II 1992)—implicitly declaring Galileo a Church theologian, whereas his aim was clearly the opposite: to free researchers from the traditional task of reconciling their ideas with holy writ.

On the more formal, juridical-theological level, Galileo refers to the hermeneutical guidelines of the Council of Trent:

> I would have doubts about the truth of this prescription, namely whether it is true that the Church obliges one to hold as articles of faith such conclusions about natural phenomena, which are characterized only by the unanimous interpretation of all the Fathers. I believe it may be that those who think in this manner may want to amplify the decrees of the Councils in favor of their own opinion.
>
> (Finocchiaro 1989, 108–109)

All in all, on the rational, theoretical, and theological levels, the *Letter to the Grand Duchess Christina*, together with Kepler's Introduction to the *Astronomia Nova*, published shortly before, marks the historical watershed between religion and science. Galileo declares the independence of the latter from the former: both seek the truth and can contribute to each other, as long as they keep a respectful distance from each other. It formalizes a process of secularization of science begun during the Renaissance, when several types of institutions other than the university began cultivating profane higher learning. Yet universities remained firmly in the hands of the clerical caste.

Not everybody agreed with Galileo, of course, and he was soon hushed by those who claimed the monopoly over science and the truth. One of them was Cardinal Robert Bellarmine (1542–1621), the most powerful theological authority in the Catholic Church (to become a saint in the 20th century). In a famous letter to a Copernican follower, Paolo Antonio Foscarini (c. 1565–1616), Bellarmine reaffirmed the subjugation of science to theology: science, he argued, cannot deal with essences; it can at the most offer mathematical models to predict phenomena, and scientists should limit themselves "to speaking suppositions and not absolutely." Hence the Copernican theory, if considered a real and not merely hypothetical description of the world, allegedly contradicts some formulations of the Holy Scripture— only then is it heretical. Bellarmine railed that the theory, as such, "is a very dangerous thing, likely not only to irritate all scholastic philosophers and theologians, but also to harm the Holy Faith by rendering Holy Scripture false" (Finocchiaro 1989, 67).

Some historians have claimed that whereas Galileo was insightful in theology, Bellarmine was insightful in philosophy of science. Looking at the controversy from today's perspective, the picture is different. Galileo went too far in claiming that science describes truth. Popper teaches that science merely quests for truth but cannot "obtain, *an ultimate explanation by essences*" (Popper 1965, 103); but this still does not mean that science is merely hypothetical, as Bellarmine requests. Bellarmine's instrumentalism, the view that a scientific theory is nothing more than an instrument for yielding correct predictions, may shelter science from a clash with religion, but is by no means a satisfactory description of the growth of science. By using different hypothetical models to explain phenomena, one has no criteria to assess which betters or furthers the progress of science. Here, the superiority of Popper's view over instrumentalism is evident. Moreover, predicting phenomena is distinctly an applied science, and thus instrumentalism leaves no room for a theoretical science.

On the theological level Bellarmine misinterpreted the guidelines of the Council of Trent (see chap. 7). His interpretation was nevertheless soon adopted by the panel of theologians at the Holy Office that in 1616 decreed Copernicanism "Foolish and absurd and in philosophy, and formally heretical" (Finocchiaro 1989, 146). Agassi (1981, 330) notes how the decision to ban Copernicanism was made with unusual speed, as though the people in charge were panic-stricken. Galileo's views threatened the church's hegemony over science and truth, and Bellarmine was making every effort to maintain it—a fatal error that not even the efforts of Pope John Paul II were able to smooth things over (see Segre 1997 and 1999). Galileo's philosophy of science may have been wrong, but Bellarmine was wrong both in philosophy of science and in theology.

For the time being, Galileo stopped disseminating the Copernican theory, but he continued to practice astronomy. In 1618 three comets, considered harbingers of the Thirty Years' War, sparked interest in, and controversy over, the nature of these unusual celestial bodies. According to Aristotelian philosophers, comets were terrestrial emanations, or rather, a "sub-lunar" atmospheric effect, because Aristotle viewed the super-lunar perfect and so unalterable, whereas comets came and went. However, a Jesuit mathematician at the Roman College, Orazio Grassi (1590–1654), rightly argued that the comets were located "above" the Moon. Strangely enough, Galileo objected to this innovative theory and had his pupil Mario Guiducci (1585–1654), Consul of the Florentine Academy, address at the Academy that comets were a sub-lunar optical phenomenon. The address was published soon thereafter. Guiducci was also member of other academies, including the *crusca* and, from 1621, the Lincean Academy. Paradoxically, it was from a Jesuit university college of medieval tradition that a voice was raised expressing opinions that were more modern than those of a secular free learned society.

All this debate took place outside the university. Galileo remained institutionally linked to the Florentine Academy and was even its Consul from 1621 to 1623. However, he did not involve himself in much of its work, an indication of the institution's decline. Apparently, he did not even flaunt the prestigious title of Consul (equivalent to that of a rector), preferring to continue presenting himself as Chief Mathematician [*Matematico primario*] and Philosopher of the Grand Duke of Tuscany (a title that conferred honor and legitimation), "super-ordinary" [*sopraordinario*] Mathematician of the *Studium* of Pisa (for which he received a salary, by order of the Grand Duke), and "Lincean."

The essence of Galileo's contribution is still debated, just as there is no consensus on how to define science itself. One thing is certain: Galileo achieved his results thanks to his willingness to break traditional and institutional frames and his appeal to his readers' own autonomous judgment

Like many illustrious predecessors, Galileo held that the appropriate way to further scientific and intellectual growth was through the dialectical confrontation of various points of view; he also understood the importance of substituting Latin with vernacular. His major work, *Dialogue Concerning the Two Chief World Systems* (1632), is an impressive popularization of science at the highest level, excelling not only in science and philosophy, but also in Italian language. It takes the form of a dialogue between three interlocutors: Salviati, who speaks for Copernicus and Galileo; Simplicio, who advocates Ptolemy and Aristotle; and Sagredo, a learned layman who acts as arbitrator of sorts, with obvious Copernican leanings. Galileo skillfully combines science, philosophy, and rhetoric into a powerful propagandist composition, diplomatically criticizing Aristotelian cosmology and arguing that the sun is the center of planetary orbits, including that of the moving earth. He also appeals to his readers' own autonomous judgment: To Simplicio's question: "But if Aristotle is to be abandoned, whom shall we have for a guide in philosophy?" Salviati answers,

> We need guides in forests and in unknown lands, but on plains and in open places only the blind need guides. It is better for such people to stay at home, but anyone with eyes in his head and his wits about him could serve as a guide for them. In saying this, I do not mean that a person should not listen to Aristotle; indeed, I applaud the reading and careful study of his works, and I reproach only those who give themselves up as slaves to him in such a way as to subscribe blindly to everything he says and take it as an inviolable decree without looking for any other reasons. This abuse carries with it another profound disorder, that other people do not try harder to comprehend the strength of his demonstrations. And what is more revolting in a public dispute, when someone is dealing with demonstrable conclusions, than to hear him interrupted by a text (often written to some quite different purpose) thrown into his teeth by an opponent? If, indeed, you wish to continue in this method of studying, then put aside the name of philosophers and call yourselves historians, or memory experts; for it is not proper that those who never philosophize should usurp the honorable title of philosopher.
>
> (Galilei 1967, 112–113)

The *Dialogue* infuriated the ecclesiastical monopolists of contemporary science to such an extent that they rushed to ban the book, put its author on trial, and isolate him by confining him to house arrest for the rest of his life.

While under house arrest, Galileo wrote his second dialogue featuring the same interlocutors: *Discourses and Mathematical Demonstrations Relating to Two New Sciences* (1638; the "two new sciences" are strength of material and kinematics). The book lays the foundations of modern kinematics. It is basically a textbook, although normally textbooks are not written as a dialogue. Galileo encourages his reader to think with his own head and consider opposing theories.

This might be the occasion for a short philosophical digression about Kuhn's debated, although popular, model of scientific growth through revolutions that grants a central importance to textbooks. Kuhn maintains that science progresses through leaps. There are two general periods in scientific activity: the "normal" period and the "revolutionary" one. In the "normal" period, in any given discipline, a "paradigm" reigns; this term, which Kuhn himself does not clearly explain, roughly indicates the scientific frame of reference at a given moment in a certain field. For example, in astronomy, for many centuries the Ptolemaic paradigm reigned, with an Aristotelian, geocentric universe composed of spheres supporting the planets. "Normal" scientific activity, in the ambit of this paradigm, consisted in solving problems (or puzzle-solving, as Kuhn calls them) by using Aristotelian or Ptolemaic premises. At a certain point, however, anomalies can arise that cannot be solved within the paradigm. For example, the discovery of the telescope revealed that not all the celestial bodies revolve around the Earth, as in the Ptolemaic-Aristotelian paradigm. Science, then, enters a "revolutionary," speculative period. A new paradigm—the heliocentric paradigm, in our example—finally supplants the old one. With it, science returns to a period of "normal" activity and soon produces a "textbook" for students to use.

Kuhn's model is, under many aspects, a faithful sociological description of *how* science often behaves, but not a recommendation of preference as to *how it should* behave. Who decides which paradigm is to be followed? According to Kuhn, the consensus of the scientific community, that is, common sense, including social, political, and irrational reasoning, rather than rational, judicious considerations. As a result scientists, scholars, and students are often guided by conformism, rather than by excellence (see appendix 2). To excel means to break a framework, not to conform to it. In many ways, Kuhn's model reflects the way the university, yesterday and today, works: it encourages conformism and at times even servility, and many centers that pretend to be of excellence are actually centers of conformism. Their "excellence" is often measured by the number of students who fail to cope with predefined "high" standards. Galileo, on the contrary, an example of true excellence, had to pay a high price for his excellence and courage. And even today, excellence can still be punished in academia, one way or another, even without official denunciation. Yet Galileo's choice inaugurated the era of new science and paved the way for his followers, some of whom achieved very important scientific results.

Galileo bequeathed the art of intellectual independence to his followers

Galileo's followers, such as Bonaventura Cavalieri (1598–1647), Torricelli, or Giovanni Alfonso Borelli (1608–1679), were independent thinkers and did not follow intellectual trends. This might be one of the reasons some had difficulty fitting in with contemporary academic institutions. Torricelli, Galileo's successor as mathematician at the Tuscan court and one of the leading mathematicians of the century—even if a priest—did not teach in a university. He was active in several academies, including the *Accademia del Disegno* and the *Accademia della Crusca*, at that time actively involved in popularizing fields that had traditionally been the domain of universities. When, in 1642, he was invited to join *Accademia della Crusca*, he received a letter from Cavalieri (another mathematical giant), which indicates how, then as now, science was viewed more as a consequence of trends or of sheer social frivolity. Cavalieri congratulates Torricelli for joining the Academy, but warns him:

> I hear that they expect physical rather than mathematical things, and perhaps they are right, for the former resemble more to the bran [*crusca*] whereas the latter is the flour—the true food and nutriment of the intellect. It is advisable to meet their expectations, and more than that, the universal expectation, that has little esteem for mathematics, unless it sees some applications.

Cavalieri scornfully tells Torricelli how to deal with this type of intellectual:

> It is therefore advisable to have ready two types of things to satisfy them all. And more than that, to satisfy the public, which decides the values of doctors and doctrines by the number of their followers, one has to avail oneself of what is more easily sold, so as to serve the public better and deceive it, or better, to kill the intellect, because the public wants to be treated this way.
>
> (*Discepoli di Galileo Galilei* 1975, 18, my translation)

Torricelli followed Cavalieri's advice: His *Lezioni accademiche,* the lectures he delivered (in Italian, of course) between 1642 and 1643 at the *Accademia della Crusca* and published in 1715, were "more physical than mathematical." Nonetheless, some of the topics were innovative. The lessons included "On the force of percussion," a controversial mathematical, physical, and philosophical argument that had been developed by Galileo, and "On the wind," an even more controversial argument tying in with Torricelli's research on atmospheric pressure. Torricelli had theorized the existence of atmospheric pressure in his explanation of the results of his famous barometric experiment, in which, in 1644, he produced the first laboratory vacuum. Aristotle had stated that vacuums could not exist, and thus, to sustain the contrary was virtually unthinkable and was basically considered a sacrilege by Aristotelian philosophers. Torricelli was clever enough to break frames while disguising his ideas.

In Italy, science went into a relative decline in the second half of the 17th century, except for niches as the natural sciences or hydrology (Maffioli 1994). All the more so in Italian universities. Leading universities such as Bologna or Padua, lost attraction. The church, typically accused of being the catalyst of the Italian scientific decline, without a doubt bears a great degree of responsibility, but it is not the only cause. The decline was more general and mainly because Italy had no central authority able to ensure the vital space science needs to develop new ideas. This space was offered by the powers that, in the meantime, had molded countries in northern Europe, such as France and England, which became centers for the development of science.

CONCLUDING REMARKS

- Modern science was generated primarily outside the university, and sometimes despite the university.
- Science requires a dialogue that goes beyond traditional barriers and clichés.
- Galileo, who can be considered the first modern scientist, was a champion in overcoming intellectual, social, and professional barriers.

BIBLIOGRAPHICAL NOTES

A vast bibliographical material is available on the scientific revolution, from Copernicus on. A useful, detailed, well documented, and interesting biography of Tycho Brahe is Victor Thoren's *The Lord of Uraniborg* (1990). Much the same can be said of John Robert Christianson's *On Tycho's Island* (2000), which concentrates more on Tycho's team of assistants at Hven.

Antonio Favaro has presented a historical overview of the University of Padua in his book *Galileo Galilei e lo studio di Padova*, published in 1883 following meticulous archival research, and reprinted in 1966. Favaro dedicates two chapters to Galileo's teachings in Padua: one to his university teaching and the other to his private lessons. Favaro was also the editor of the "National Edition" of Galileo's Works. Thanks to its thoroughness, this edition is probably the best collection of works by a scientist ever published.

Karl Popper deals with Galileo between instrumentalism and essentialism in "Three Views Concerning Human Knowledge," first published in 1956 and republished in his *Conjectures and Refutations*, pp. 97–119.

The context surrounding Galileo's controversy regarding comets is described, inter alia, in Pietro Redondi's classic *Galileo Heretic*, first published in Italian in 1983. According to Redondi, the reasons for the "Galileo affair" are in the comets' controversy and the subsequent publication of the *Il Saggiatore* (1623, in English: *The Assayer*), in which Galileo supposedly expressed himself heretically—not regarding astronomy, but rather regarding atomism. Redondi's thesis has been much criticized, but the book remains exciting to read and authoritative enough, presenting the social background of Galileo's works.

For Galileo and his followers, see my *In the Wake of Galileo* (Segre 1991a).

NOTES

1. Olschki 1965c, 140, wrongly claims that mathematics was not taught at the University of Pisa at the time Galileo was studying there in 1584–85. The same list (see *OG* 19, 34) attests that mathematics was taught by one Filippo Fantoni.
2. From Pietro Bucci, *Le coronationi di Polonia, et di Francia* (see n. 1 in Favaro's book, same page). My translation.
3. http://www.fisicamente.net/FISICA/LEOPERE10.htm#_ftnref365, accessed January 1, 2014.

10 Science Develops Outside "Academia"

Early modern philosophy of science, inaugurated by Francis Bacon, despite shortcomings, criticized Academia and paved the way toward a more open scientific society[1]

What was that new knowledge—science—which emerged in the 17th century, and is today monopolized by universities? The question was put as early as the beginning of the century by Francis Bacon (1561–1626), Lord Chancellor of England and the first great philosopher of science. Bacon was not an academic in the current sense of university membership. He studied, but did not graduate, at the University of Cambridge, and then attended Gray's Inn in London, dedicated to legal study. He pursued a political career and made his contributions in philosophy far from the university. Bacon was the first philosopher to notice that knowledge grows and to openly recognize and highlight the utilitarian role of science, inter alia, as a source of power. He wondered how knowledge grows and what factors may foster or hinder its growth.

Bacon's work is of central importance to understanding developments in modern higher learning. Besides studying knowledge and its growth, Bacon deeply influenced the way science is perceived and studied up to the present day. Bacon rejected tradition, prejudice, and dogmatic thinking, as well as the confusion between science and theology. Like his contemporaries Galileo and Kepler, Bacon advocated for a separation between theology and science. He rejected, of course, Aristotelian philosophy, at times in flagrant contradiction to reality, urging instead to think objectively, to dismiss prejudices and preconceptions, and to observe reality carefully and systematically, trusting the senses. He also criticized the contemporary university, and some of this criticism is still pertinent today: it is excessively oriented toward the teaching of some professions and hostile to innovation; it neglects universal principles and discourages creativity; its professors are underpaid, and there is no exchange of knowledge among institutions. In his *Novum Organum* ("New instrument" [of science], 1620, in contrast to Aristotle's *Organon*), he says,

In the customs and institutions of schools, academies, colleges, and similar bodies destined for the abode of learned men and the cultivation of learning, everything is found adverse to the progress of science. For the lectures and exercises there are so ordered that to think or speculate on anything out of the common way can hardly occur to any man. And if one or two have the boldness to use any liberty of judgment, they must undertake the task all by themselves; they can have no advantage from the company of others. And if they can endure this also, they will find their industry and largeness of mind no slight hindrance to their fortune. For the studies of men in these places are confined and as it were imprisoned in the writings of certain authors, from whom if any man dissent he is straightway arraigned as a turbulent person and an innovator.[2]

On the other hand, Bacon produced a complex, and not entirely clear, inductive heuristic for the growth of knowledge (i.e., a procedure of scientific investigation whereby one generalizes from details), and today it is still generally wrongly believed, at least on the popular level, that science is observation laden and grows by means of induction. According to Bacon, knowledge accumulates (rather than evolves as Popper contends) and must be gathered and classified—a method still to be found in university studies, often burdening students with details. To be sure, Bacon was an idealist: he preached humility in research and was most probably not aware that his views of science might foster authoritarianism and at times arrogance. In schools and universities, however, the pedantry of the teacher or the scientist causes discomfort to the student (a practice, as seen, already criticized by ancient Jewish sages). Not infrequently, students are taught to collect and memorize details, with little attention to the context. In other words, they are provided with answers to unasked questions. Quantity, as enumerated in the number of pages to study or write and publish, is preferred to quality and the originality of the ideas. Alexandre Koyré (1892–1964), one of the greatest historians of science ever, regards Bacon only as "the announcer, the *buccinator*,[3] of modern science, not one of its creators" (Koyré 1992, 17, n.1). According to Koyré, "The 'activism' of modern science, so well noticed—*scientia activa, operativa*—and so deeply misinterpreted by Bacon, is only the counterpart of its theoretical development" (Koyré 1992, 90).

Bacon's philosophy of science bears new basic philosophical, educational, and implicitly social problems, foremost being the "problem of induction," as illustrated by the classical example of the white swans: observing a large number of white swans can never verify that all swans are white. In other words, one cannot make a universal claim based on a finite number of observations because only one observed counterexample serves as refutation. Inductive methodology, however, spread and developed into an influential philosophical current: empiricism, whose most prominent representatives were John Locke (1632–1704), George Berkeley (1685–1753), and David Hume (1711–1776). They all attempted, unsuccessfully, to solve the problem of induction. The problem was finally solved by Popper. His heuristic runs as follows: Begin your research by posing a question to nature. Conjecture an answer even without prior observation. If the said answer is testable empirically, then you are doing science, and if it works, you have a theory that you may use, as long as it works. For, if/when the theory is empirically refuted, it is time to conjecture a better theory and repeat the process. This is how science evolves, possibly endlessly, nearing truth but never reaching it. All this, of course, is better performed in a society that encourages criticism, that is, an "open society," rather than in a dogmatic society, that is, a "closed society."

While attempting to get rid of old dogmas, Bacon unintentionally paved the way to new ones, ultimately leading to the still current positivistic belief that science is omnipotent. "Positivism," to be sure, is not easy to define. According to the *Encyclopedia Britannica* (1960) the term

> May be applied to any system that confines itself to the data of experi-
> ence and excludes a priori or metaphysical speculations. In this sense
> the term is commonly applied to the empirical philosophers, although
> in fact reservations ought to be made (John Locke and David Hume
> accept mathematics, Locke and George Berkeley accept a knowledge of
> the soul and of God, on nonempirical grounds).

Bacon was no doubt a positivist, if not the founder of modern positivism, and his philosophy influenced pedagogy as well.

The tension between the Baconian philosophy and traditional university curricula, however, sparked a broad-ranging debate in England, as in the rest of Europe, and a call for reform. Bacon's follower, intelligencer (i.e., disseminator of news), pedagogue, and polymath Samuel Hartlib (c. 1600–1662), who, just like the former, studied in Cambridge but did not graduate, in "Some Proposals towards the Advancement of Learning" (1563) complained that universities are of

> Monkish Constitutions and Customs, whereby the members of these
> Societies are obliged to none other care, but to live within themselves,
> and to themselves a dronish life, is a shew of upholding the name of
> learning, without any engagement towards the publick concernements
> of the Commonwealth of learning.
>
> (Webster 1970, 191)

Hartlib had set up a correspondence network covering Western and Central Europe. In concert with the Moravian pedagogue Jan Amos Komenský (1592–1670, also known as Comenius), Hartlib upheld the importance of a universal dissemination of education, making it more digestible, and encouraging direct contact between the student and his environment. Yet Comenius's pedagogy, too—although progressive and welcome—encouraged, in the wake of Bacon, the cumulative concept of knowledge, with its defects. Locke, among the empiricists, was well aware of the damage an authoritarian teaching may cause, and endeavored that teachers act minimally, with no initiative, to avoid possible damage: they should avoid deciding the agenda of pupils, as these depend on interest. They may liven interest but still stay behind it.

Despite drawbacks, therefore, one cannot ignore the fresh modus of thinking conveyed by Bacon, Comenius, Locke, and other contemporary philosophers and scientists. Bacon would have liked to see science cultivated by wealthy institutions or well-organized centers that employ researchers in team work. His *New Atlantis* (written around 1624 and published in English in 1626—the year after Bacon's death—and in Latin in 1638) describes an island inhabited by a utopian society owing its quality of life to science and technology. At its center is "Salomon's House," a research institution that, according to some scholars, is the imaginary prototype of the modern scientific academy, with lay researchers (there is, at least, no mention of clerics being involved) having well-defined roles and appropriate working conditions. New Atlantis and Salomon's House, nevertheless, keep their discoveries secret from the rest of the world, even though total openness reigns on the island.

Times were nevertheless changing. As Eamon (1991) suggests, various factors paved the way to the evolution in communications, the opening of society in the Popperian sense, and the spread of science: the advent of printing and its outcomes, above all; the widespread association between progress and the need to keep up; and the weakening of secretive institutions, such as guilds and other types of associations. This led, inter alia, to a greater awareness of the right to intellectual property and to the creation and consolidation of guarantees through patents and copyrights as an alternative to secrecy. In this context the description of lay scientists working in an atmosphere of open communication presented in Bacon's *New Atlantis* may have had a fundamental impact. But it also contains some hints towards a new type of secular religion, with its rites, saints, and clerics: in presenting Solomon's house, its "father" (i.e., president) says,

> For our ordinances and rites: we have two very long and fair galleries: in one of these we place patterns and samples of all manner of the more rare and excellent inventions: in the other we place the statuas of all principal inventors. There we have the statue of your Columbus, that discovered the West Indies: also the inventor of ships: your monk that was the inventor of ordnance and of gunpowder: the inventor of music: the inventor of letters: the inventor of printing: the inventor of observations of astronomy: the inventor of works in metal: the inventor of glass: the inventor of silk of the worm: the inventor of wine: the inventor of corn and bread: the inventor of sugars: and all these by more certain tradition than you have. Then have we divers inventors of our

own, of excellent works; which since you have not seen, it were too long to make descriptions of them; and besides, in the right understanding of those descriptions you might easily err. For upon every invention of value, we erect a statua to the inventor, and give him a liberal and honourable reward. These statua's are some of brass; some of marble and touch-stone; some of cedar and other special woods gilt and adorned; some of iron; some of silver; some of gold.

We have certain hymns and services, which we say daily, of Laud and thanks to God for his marvellous works: and forms of prayers, imploring his aid and blessing for the illumination of our labours, and the turning of them into good and holy uses.

Bacon (2002, 487–488)

Modern science was created by amateur gentlemen scientists who found their legitimization in chartered scientific societies

In various countries in Europe many new institutions appeared, such as botanical gardens, astronomical observatories, and hospitals, that furthered science much more than universities did. In France, where the Baconian spirit was applied more assiduously than in England, Louis XIII (r. 1610–1643) and Cardinal Richelieu (1585–1643), following the initiative of Louis's physician, Guy de La Brosse (1586–1641), created the *Jardin royal des plantes médicinales*, also known as *Jardin du Roi*. It was inaugurated in 1640 and offered free public lessons in French, and not in Latin, on innovative topics such as the circulation of the blood. It sparked the opposition of the medical faculty of Paris's university, the Sorbonne, which prohibited it from conferring titles. The *Jardin du Roi* had many functions, from the scientific endeavor of botanical research to the political responsibility of taking advantage of research in the interest of the country, including the glorification of the king, and the assimilation and exploitation of the many plant species found in the colonies for use in France, thereby indirectly justifying colonialism.

Italy, more specifically, Tuscany, the cradle of the modern learned societies, is also the birthplace of the first court-sponsored scientific academy in the sense of "hard" sciences. During the decade 1657–1667, the *Accademia del Cimento* (*cimento* means "experiment," or "trial"), although not chartered, was active at the Tuscan court. It was led by Prince Leopold de' Medici (1617–1675), the brother of the Tuscan grand duke, and by Italy's greatest contemporary scientist, Borelli. One other leading member was Viviani, Galileo's "last pupil" (as he called himself) and biographer. Viviani was not a cleric, and I did not find any evidence that Borelli was one; in any case, the academy was surely not a clerical institution. The other members of the academy were physicians, philosophers, or men of letters. None of the members received a salary for his work within the academy. Borelli earned his life as professor of mathematics at the University of Pisa—a position that did not satisfy him and that he ultimately abandoned. Viviani was employed by the court as engineer, although in 1664 he was granted a small pension by King Louis XIV (r. 1654–1715), and in 1666 the Tuscan grand duke, Ferdinand II (r. 1621–1670), relieved him of his public obligations, granting him a salary to enable him to concentrate on his studies. The initial purpose of the *Accademia del Cimento* was purely scientific, and the prince put at the disposal of its members an unprecedented array of instruments and means to carry out projects that could have become the vanguard of contemporary science. It is a mystery why it survived so briefly. Paolo Galluzzi (1981) blames the oppressive interference by the prince, who was in turn conditioned by the complex political and cultural circumstances in the second half of the century. Personal tensions among the members' participants led the prince to impose a self-censure on the publication of some important results, and ultimately, the academy dissolved. The only published results were the *Saggi di naturali esperienze* (1667), an impressive volume with descriptions and illustrations of experiments that hardly represents the true work of the academy, which is described in dozens of hand-written volumes in the Central National Library of Florence still waiting to be published and studied. In any case, the *Accademia del Cimento*, although still with some elements of secrecy, marked the dawn of the golden age of court-sponsored scientific academies. Next to come was the celebrated Royal Society of London.

The Royal Society of London, which still exists today, was more successful, and its origin and history have been subject to many studies. As early as the 1640s, a heterogeneous group of English intellectuals, noblemen, clergymen, officials, and instrument makers met privately in London or Oxford to discuss the progress of science. In London they met, inter alia, at Gresham College. In 1660 they founded the "Royal Society," committed to the physical-mathematical and experimental science. Its motto, *"Nullius in verba,"* according to its first historian, Thomas Sprat (1667), was inspired by Baconian thinking. Sprat alluded more to the methodological aspect of the Royal Society because on an institutional level it was quite different from Salomon's House. Boyle, who, among other, coined the term *invisible college* (sometimes referred to as Hartlib's circle), was among its prominent founders and its intellectual leader.

Boyle, like Bacon, was not an academic: he was a wealthy nobleman who did not attend a university. At the age of less than nine years, his rich father sent him to Eaton to study under Sir Henry Wotton (1568–1639), writer and provost of the college. Wotton had also been a diplomat (he had been, inter alia, English ambassador to Tuscany), a member of the House of Commons, and, more in general, a dilettante lover of wisdom who influenced Boyle politically and intellectually; one of Boyle's leading purposes, as far as the present work is concerned, was

To fulfill Bacon's dream as he understood it and create a society of amateur scientists. He may have learned from Wotton's example how unhappy is the result of purely dilettante research in the purely Baconian fashion and how powerful the amateur's love of wisdom can be—especially when supported with a little guidance. One way or another, Boyle deliberately played the dilettante and he urged others to join him, adding to the amateurism he propagated a few simple rules that sufficed to turn the collection of facts, apparently similar to that of Bacon, into scientific material—at least partly so.

(Agassi 2013, 161–162)

A society of amateur scientists implies, of course, scientists not belonging necessarily to a caste of clerics. Here is one basic social contribution of the Royal Society inspired by Boyle and often neglected. "Boyle advocated a view that he ascribed to Philo Judeus: research is the worship that befits natural religion" (Agassi 2013, 128). Other members of the Royal Society went even further: they wanted Boyle "to establish a secular college and he staunchly refused" (Agassi 2013, 129). As a result, Sprat relates that the Royal Society

Is abundantly cautious, not to intermeddle in *Spiritual things*: But that being only a general plea, and the question not lying so much on what they do at present, as upon the probable effects of their Enterprise; I will bring it to the test through the chief Parts of *Christianity*; and shew that it will be found as much avers from Atheism, in its issue and consequences, as it was in its original purpose.

(Sprat 1667, 347)

Of course, every founder and member of the Royal Society had his own view of the relation between religion and science. "For Sprat, the exclusion of discussions of 'divinity' was not motivated by a belief in the irrelevance to experimental science of general religious concerns, but by the desire to avoid unnecessary and pointless debates about technical and indifferent points of theology and ritual" (Harrison 2010, 5). However, together with a new type of academic association, the seeds of a new, secular, religion were laid, starting from the attempt to create a social movement around science as a faith of intelligent modern people, and culminating in 19th-century positivism and leaving traces up unto the present day.

Among Boyle's many contributions to science and philosophy, his urging that every scientific experiment be repeatable and that scientific activity be made public was a further step towards an open scientific community. Yet Boyle is famous mainly for the physical law that bears his name, correlating the volume (V), the pressure (P), and the absolute temperature (T) of any given gas: $P\,V = R\,T$, where R is a constant. This, however, is a modern formulation of the law. As Agassi (1977, 198–99) remarks, Boyle only said that the pressure of a gas is proportional to the density.

One other founder of the Royal Society was John Wallis (1616–1703), a cleric whom Martha Ornstein (1879–1915) considers "the greatest pre-Newtonian English mathematician, and through his important work, *The Arithmetic of Infinitesimals*, paved the way to Newton's discovery of the calculus. (Ornstein 1963, 94). Wallis, who had studied mathematics and medicine in Cambridge, "had acquired a thorough knowledge of contemporaneous continental work in geometry, through independent study," and "had to leave Cambridge because 'that study had died out there'" (Ornstein 1963, 94, also quoting Mullinger 1911, 462). Yet Wallis was appointed in 1649, by parliamentary order, Savilian Professor of Geometry in Oxford. The chair had been established in 1619 by the mathematician Sir Henry Savile (1549–1622), Warden of Merton College.

In 1662 the Royal Society of London received a first Royal Charter form King Charles II (r. 1660–1685). The Charter authorized the Society, among other,

> to enjoy mutual intelligence and knowledge with all and all manner of strangers and foreigners, whether private or collegiate, corporate or politic, without any molestation, interruption, or disturbance whatsoever: Provided nevertheless, that this our indulgence, so granted as it is aforesaid, be not extended to further use than the particular benefit and interest of the aforesaid Royal Society in matters or things philosophical, mathematical, or mechanical.[4]

The endeavor for openness speaks for itself. Moreover, the restriction to "philosophical, mathematical, or mechanical" matters may point out a further attempt to separate science and religion. A second charter came a year later, referring to it as "The Royal Society of London for Improving Natural Knowledge"; a third charter was granted in 1669, and the Society gradually became Europe's most important scientific institution. Isaac Newton (1643–1727) joined the Society as a fellow in 1671, and the Society sponsored the publication of his *Philosophiæ Naturalis Principia Mathematica* (1687, in short *Principia*), one of the basic and most important works in the history of science. Newton presided over the Royal Society from 1703 till his death in 1727.

The Royal Society sought to reform scientific learning for the good of mankind; it promoted a line of thought that criticized the blind faith that universities placed in the classical texts, and, in the wake of Bacon, it praised empiricism. Here is Sprat's description of its work, whether accurate, or embellished to fit the Baconian Recipe:

> Their manner of gathering, and dispersing *Queries* is this. First they require some of their particular Fellows, to examine all Treatises, and Descriptions, of the Natural, and Artificial productions of those Countries, in which they would be inform'd. At the same time, they employ

others to discourse with the Seamen, Travellers, Tradesmen, and Merchants, who are likely to give them the best light. Out of this united Intelligence from Men and Books, they compose a Body of Questions, concerning all the observable things of those places. These Papers being produc'd in their weekly Assemblies, are augmented, or contracted, as they see occasion. And then the Fellows themselves are wont to undertake their distribution into all Quarters, according as they have the convenience of correspondence: of this kind I will here reckon up some of the Principal, whose Particular heads are free to all, that shall desire Copies of them for their Direction.

They have compos'd Queries, and Directions, what things are needful to be observ'd, in order to the making of a Natural History in general: what are to be taken notice of towards a perfect History of the Air, and Atmosphere, and Weather: what is to be observ'd in the production, growth, advancing, or transforming of Vegetables: what particulars are requisite, for collecting a compleat History of the Agriculture, which is us'd in several parts of this Nation.

They have prescrib'd axact Inquiries, and given punctual Advice for the tryal of Experiments of rarefaction, refraction, and condensation: concerning the cause, and manner of the Petrification of Wood: of the Loadstone: of the Parts of Anatomy, that are yet imperfect: of Injections into the Blood of Animals; and Transfusing the blood of one Animal into another: of Currents: of the ebbing, and flowing of the Sea: of the kinds, and manner of the feeding of Oysters: of the Wonders, and Curiosities observable in deep Mines.

(Sprat 1667, 155–156)

A Baconian portrait indeed, although, as Agassi remarks, "The Society did more as an instrument for the advancement of learning: contrary to its own ideology, it encouraged the development of hypotheses and controversies, and contributions of individual thinkers as individuals head and shoulders above their colleagues" (Agassi 1981, 370).

Among its many activities the Royal Society supported a research laboratory at Gresham College where experiments suggested by its fellows were carried out. Boyle's assistant, Robert Hooke (1627–1691)—known for the law that bears his name saying that the power of a spring is proportional to its tension—was appointed curator of the experiments of the Society. Above all, the Royal Society gave its members legitimation and facilitated the exchange of ideas. In 1665 it began publishing the *Philosophical Transactions*, edited by the society's "second secretary," the German-born Henry Oldenburg (c. 1619–1677), at his own personal expenses. The "first secretary" was the clergymen and scientific popularizer John Wilkins (1614–1672). Oldenburg had traveled around Europe and, like Boyle and Wilkins, was not a member of a university. In the 1640s he spent some time in Holland and "found the intellectual atmosphere at the University of Utrecht uncongenial"

(Boas Hall 1965, 277). Oldenburg was much of a catalyst within the scientific community, pressuring scientists to divulge their projects or discoveries, or communicating them through his broad correspondence "providing an invaluable intelligence network without which, as contemporaries were only too aware, effort was easily duplicated or findings lost" (Hunter 1981, 49). The *Philosophical Transactions* enabled quick and detailed publications and set procedures of acknowledgment. One of its first major publications, in 1665, was Hooke's *Micrographia,* presenting his microscopical observations. Fortunately, yesterday as today, many scientific journals and books are published relatively independently from universities. Yet, they are often published or rejected on the basis of (debated) peer reviews and more committed to conformism rather than to excellence. Scientific journals are today rated by the number of (non-conformist) articles they reject, just as universities are rated by the number of (non-conformist) students they reject or fail—hardly an encouragement of excellence. As Agassi (1990a, 171) scornfully remarks concerning peer reviewers, they "have to learn to behave in the style of big corporations and learn to relate properly to power brokers and often enough lose their self-respect."

All the social-institutional opening took place within a broader social context of the Reformation, which favored utilitarianism and piety, factors that were foreign to the contemplative nature of the universities at the time. Merton claims that it was the spirit of Puritan Ethos that provided the impetus for scientific and technological development. As far as the universities and the Royal Society are concerned, he comments,

> It is well known that the universities were the seats of conservatism and virtual neglect of science, rather than the nurseries of the new philosophy. It was the learned society which effected the association and social interaction of scientists with such signal results. The *Philosophical Transactions* and similar journals largely did away with the previously prevailing and unsatisfactory mode of communicating new scientific ideas through personal correspondence.
>
> (Merton 2001, 103)

Undoubtedly, from the 17th century onward, Protestant societies underwent a lively scientific and technological development. Nonetheless, if one considers details, Merton's thesis proves shaky. Why did Protestant universities not follow this vein of scientific and technological progress? Can the Catholic world's important contributions to science, such as that of Galileo and his followers or those of Catholic French scientists and intellectuals in the 17th and 18th centuries, be ignored? This brings up a few fundamental historiographical questions: Why does the scientific community give such importance to a thesis like Merton's? Could it be that irrational factors such as intellectual trends or ideologies played a role? Which factors determine the consensus of the scientific community so that, according to Kuhn, they determine new "scientific paradigms"? How much attention is paid to rational and valid criticism, even though it might prove painful to some? I cannot repeat too often that the encouragement of controversy and criticism is one of the vital conditions for the growth of science and knowledge.

The Royal Society, which was much more of a club, gave impetus to this new element of communication and openness; despite its meager financial means, it became a meeting point for the European scientific community, a place where scientists could openly debate their personal opinions through direct confrontations, correspondence, or ad hoc publications. Leading men of science from overseas joined the society. Among them, physicist, mathematician, and astronomer Christiaan Huygens (1629–1695), who proposed, inter alia, that Saturn was surrounded by a ring and later theorized the wave theory of light; or leading anatomist Marcello Malpighi (1628–1694), Borelli's pupil, who performed extraordinary microscopic observations in plants and animals; and Gottfried Wilhelm Leibniz (1646–1716).

The Royal Society was obviously more "open" (philosophically speaking) than most universities and managed to overcome many of the factors that hindered the development of science in universities. Its main limitation was the lack of funds. This drawback, as we shall soon see, was overcome, at least temporarily, by the French *Académie royale des sciences*. But more presently, in England, science remained a respectable leisure time occupation for country gentlemen and townsman of substance. Charles Watson-Wentworth, 2nd Marquess of Rockingham (1730–1782), twice Prime Minister of England, for example, entertained himself by conducting experiments on the production of oil from tar. The various philosophical clubs and societies, such as the Lunar society of Birmingham (founded in 1766 by Erasmus Darwin, 1731–1802—a physician, grandfather of Charles Darwin, 1809–1882) or the Literary and Philosophical Society of Manchester (founded in 1781), offered lectures on topics such as electricity, hydrostatics, gravitation, mechanics, and optics, performing experimental demonstrations to fashionable and enthusiastic audiences. But the universities shared little of this enthusiasm.[5]

Developments similar to those occurring in England occurred in France. Like the Royal Society of London, the *Académie des Sciences* had a period of incubation of several decades. As L.W.B. Brockliss (1987, 446–447) remarks, during the 17th and 18th centuries, many of the leading scientists did have an academic position, but specifically as teachers in the new and well-endowed State institutions: the *Collège Royal* and the *Jardin du Roi*. In addition, much like in England, informal groups of intellectuals met to discuss literary, political, and scientific topics. One particularly interesting figure was that of Marin Mersenne (1588–1648), a friar of the Order of the Minims and an admirer of Bacon. He, too, corresponded with a great number of intellectuals in France and abroad, propagated and criticized their scientific ideas, and became a reference point—a true one-man institution. Moreover, a group of intellectuals interested in philosophy and mathematics would meet at his house. René Descartes (1596–1650) and Pierre Gassendi (1592–1664) were among the prominent philosophers belonging to Mersenne's circle and not to a university establishment. Gassendi was a priest, an atomist philosopher, and a lecturer at the *Collège royale.*

Descartes's contribution to philosophy and science can hardly be outlined. In his *Discours de la méthode (Discourse on Method,* 1637) he drew a rational, deductive heuristic. In *La Géométrie (Geometry),* published in the same year, he formulated the analytical geometry. In his *Principia Philosophiæ (Principles of Philosophy,* 1644) he developed a cosmological theory of vortices filling the universe and moving the celestial bodies. This is only part of Descartes's contributions. As Agassi notes, Descartes and Bacon were the first radicals because they demanded to start afresh. "Radicals are those who try to push it to the limit," and "radicalism says that if you go about things the right way, if you first get rid of all the rubbish that was your cultural and/or your scientific heritage, you are on the way to salvation" (Agassi 1990b, 157 and 158). Yet unlike Bacon, Descartes was a deductivist, and his method can be summarized as follows: get rid of all the rubbish, find an axiom to begin with, make sure it is true, and deduce from it all knowledge. The big difficulty remains to determine the truth of an axiom; according to Popper this is impossible: science is not truth, but the pursuit of it. The Cartesian philosophy, by the way, was rejected at Oxford and attacked by its bishop, Samuel Parker (1640–1688).

One other circle devoted to medicine, mathematics, liberal arts, and mechanics was the "Montmor Academy," meeting at the residence of the wealthy Parisian Henri Louis Habert de Montmor (c. 1600–1679). In 1653 Gassendi moved to live there, attracting around him leading mathematicians and physicians, such as Girard Desargue (1593–1662), Gilles Personne de Roberval (1602–1675, also a lecturer of mathematics at the *Collège Royale*), and Blaise Pascal (1623–1662), none of whom belonged to the university establishment. These and other groups may have contributed to the foundation of the *Académie Royale des sciences* (1666) under Louis XIV at the suggestion of Jean-Baptiste Colbert (1619–1683).

The first "professional" scientists appeared in the French Académie royale des sciences, *although scientific practice can hardly be called a profession*

Unlike the Royal Society, the *Académie Royale des Sciences* was financed and organized directly by the crown; its members were salaried and enjoyed many facilitations offered by the court, and this made the academy subject to the king. James McClellan (1985, 13) distinguishes between "societies"—which were learned and scientific and spread in Britain, the American Colonies, and Holland—and continental "academies," adding yet one more meaning, albeit more modern, to the term "academy." Whereas *societies* were a kind of club, salaried *academies* were more rigid and hierarchical.

Conferring salaries to scientists was the first step toward transforming science from an amateurish activity into a profession, or better said, a pseudo-profession. Can scientific activity be considered a "profession," that is, an activity that confers a (preferably honorable) living? Yes, according to sociologist of science Joseph Ben-David (1920–1986), at least from the 18th century onwards. His thesis is today generally accepted but debatable. To be sure, the members of the *Académie des Sciences* may be considered as "professional" scientists, although many of them had other resources; one can argue that science—whether "theoretical" or "experimental"— is only rarely a profession. Applied science is undoubtedly a profession, but in this case it is, in the end, engineering. To paraphrase Roger Hahn (1971, 40), beyond economic and educational impediments, there were and still are deep-seated psychological motives that now as then prevent the true man of learning from identifying with either the professional or the craftsman. The typical professional represents limitations rejected by a genuine man of science. For example, priests and professors were and are inevitably tied to received authority; lawyers have always been famed for their skill in distorting truth for the benefit of their clients. Indeed, Einstein was of the opinion that a scientist should earn his living from a "cobbler's job" (Frank 1953, 110). A genuine scientist or any individual that creates excellence, whether in the "hard" or in the "soft" sciences, cannot be confined within a closed ideological, religious, or social frame. To put it in Popper's terms, he or she does not fit in a closed society. The *Académie Royale des Sciences* was nevertheless open enough to become the leading scientific institution in Europe.

Like the Royal Society, the *Académie des Sciences* was open to, and welcomed, foreign scientists, recruiting some of the best minds in Europe. Among them were Huygens, who lead the academy, and the outstanding Italian telescope maker Giandomenico Cassini (1625–1712). One project of the *Académie* was to supervise the building and the function of the Paris observatory, whose work inaugurated the era of astronomical precision. Its building took five years and was completed in 1672, although astronomers already worked there before its inauguration: Cassini moved to the observatory in 1671. Among Cassini's findings are four satellites of Saturn, the shadow of Saturn's ring on the body of the satellite, and the gap in Saturn's ring. Other projects carried out at the Observatory were the discovery and measurement of the annual displacement of the pole star; the determination of the parallax of Mars and indirectly of the dimensions of the solar system; the discovery of the flattening of Jupiter; the discovery of the finitude, and the determination, of the speed of light, suggested by Ole Rømer (1644–1710), as an explanation to Cassini's observed discrepancy between the predicted and the observed motion of a Jupiter satellite; and the accurate mapping of the moon. The Observatory and the Academy contributed both to state projects and general projects, such as meridional surveys within and outside of France.[6]

The changing demands of the society were among the main causes of the decline of the *Académie*, as well as other academies. They are well described by Hahn:

> With their gaze fixed on the progress of their profession, most scientists failed to consider how changes in eighteenth-century society might bring into question the very premises upon which their institution was based. Nevertheless, in at least three fundamental and interrelated ways the position of the Academy was slowly being undermined. Utilitarian demands placed upon the royal institution multiplied at such a pace that the delicate balance between its research and consultative functions was jeopardized. A growing egalitarian sentiment fostered dissatisfaction with the hierarchical organization and elitist principles cherished by the Academy. Romanticism challenged the assumptions of the classical age, including the very rationale behind both the academic system and the kind of collective activity in which the academicians were engaged.
>
> (Hahn 1971, 116)

For the time being, however, the *Académie Royale des Sciences*, and, more in general, the French scientific institutions inspired the founding of analogous German institutions. The oldest was created as early as in 1652, when four physicians from Schweinfurt on the Rhine, in Franconia, founded the *Collegium Naturae Curiosorum*, dedicated to medical treatment. In 1687, it received the protection of Emperor Leopold I (emperor between 1658 and 1705) and adopted the name *Sacri Romani Imperii Academia Caesareo-Leopoldina Naturae Curiosorum*, abbreviated as the "Leopoldina," still existing. Imperial recognition included privileges (sometimes odd), such as raising the president and director of its publication, *Miscellanea*, "to the dignity of count palatinate, with the right of legitimizing illegal children, sanctioning adoptions, etc." (Ornstein 1963, 173). The academies began to follow the path of late medieval universities, meddling with social issues, and this perhaps foretold their decline, even though they remained a pillar of scientific progress until the end of the 18th century.

In 1700, on the initiative of Leibniz and with the support of Elector (and from 1701 till his death, in 1713, king) of Brandenburg-Prussia, Frederick I, the *Churfürstlich-Brandenburgische Societät der Wissenschaften* ("Electoral Brandenburg Society of Sciences and Humanities") was founded. Leibniz became its president and envisioned an academy that was as free as possible, even from state control. The academy was given definitive recognition in 1711. In 1740, immediately after becoming king of Prussia, Frederick II ("the Great," r. 1740–1786), grandson of Frederick I, reorganized the society, inter alia, following suggestions by Voltaire (1694–1778). In 1744, it was merged with the *Société Littéraire de Berlin* (Literary Society of Berlin, founded in 1743—the name indicates French influence) and took on the name *Königliche Preussische Akademie der Wissenschaften* (Royal Prussian

Academy of Sciences). The Prussian Academy, besides the relatively free environment envisioned by Leibniz, had access to many facilities: scientific instruments, an anatomical theatre, a botanical garden, and natural history collections. These became precious when, at the beginning of the following century, the "Free" University of Berlin was founded, an event that marked a turning point in the history of universities. We will deal with it in the next chapter.

During the course of the 18th century, many other royal or imperial scientific academies—such as the one Peter the Great (r. 1682–1725) established in 1724 in Saint Petersburg—were founded and became the institutional fulcrums of scientific research.

As opposed to universities, learned societies with their dilettante membership were a typical example of what Aristotle would have considered "learning and education which we must study merely with a view to leisure spent in intellectual activity, and these are to be valued for their own sake" (see chapter 3). The contemporary leading figures in science might have studied at a university, but their contributions were made outside its walls. Newton, for example, laid the basis of his scientific discoveries at home when the University of Cambridge—where he was enrolled as a student at Trinity College—had to close because of an outbreak of the plague. In the space of less than a year Newton managed, all by himself, to assimilate the mathematical achievement of a century and placed himself at the forefront of European mathematics and science (Westfall 1980, 144). True, in 1669 Newton was appointed Lucasian Professor at Cambridge. The Lucasian Chair of Mathematics, still existing, had been established six years earlier as a gift from Henry Lucas (c. 1610–1663), once the university's Member of Parliament. Newton was, apparently, not the best teacher. In any case, by then, "Mortal illness ate at the life both of the university and of Trinity [. . .] a precipitous decline, which reduced the university in the space of two decades nearly to half its former size, was about to set in early the 1670s. University institutions were increasingly shams" (Westfall 1980, 183). Newton, who relied mainly on the Royal Society, left Cambridge in 1696 to become Warden of the Mint, that is, responsible for investigating cases of counterfeiting.

> After a residence in Trinity of thirty-five years, he contrived to depart, bag and baggage, in less than a month, part of which he spent in London. While he continued to hold both fellowship and chair, and to enjoy their incomes, for another five years, he returned only once for half a week to visit. As far as we know, he wrote not a single letter back to any acquaintance made during his stay.
>
> (Westfall 1980, 550)

Following are a few additional examples of leading men of science working outside the university establishment: Paradigmatic is Henri-Louis

Duhamel du Monceau (1700–1782), several time president of the *Académie Royale des Sciences*, who, among other, experimented and developed, in his own estate, new agronomical methods. Leonhard Euler (1707–1783), studied philosophy at the University of Basel, but became a leading mathematician, working first at the Imperial Russian Academy of Sciences in Saint Petersburg, then, for many years, at the Berlin Academy, and eventually returning to Saint Petersburg. In 1739, Euler's peer, Georges-Louis Leclerc, Count of Buffon (1707–1788), became the intendant (director) of the *Jardin du Roi* and remained in this position for half a century, turning it into one of the most important scientific institutions in Europe. Buffon, too, did not belong to the university establishment: he had studied at the University in Anger (western France) but had to leave the university after having killed an adversary in a duel. Henry Cavendish (1731–1818) studied physics and mathematics at Cambridge but left the university without a degree. A wealthy man and a member of the Royal Society, although shy in temperament, he conducted his research on his own, obtaining remarkable results, among them, isolating hydrogen, and synthesizing water, measuring gravitational constant in laboratory, and deducing the average density of the Earth. Along with Charles Augustin de Coulomb (1736–1806), Cavendish set the basis of electrostatics; Coulomb himself was a military engineer who became a member of the *Académie des Sciences*. Joseph-Louis Lagrange (1736–1813), who, among other, re-formulated Newtonian mechanics in simpler formulas in his monumental *Mécanique analytique* (1788), was supposed to study law at the University of Turin but never graduated and was an autodidactic mathematician. At the age of 19 he was appointed teacher of the Theory and Practice of Artillery at the Piemondtese Royal Military Academy, and in 1766 Frederick II of Prussia invited him to join the Berlin Academy as "the greatest mathematician in Europe." In 1786, following Frederick's death, Lagrange accepted the offer of Louis XVI to move to Paris and became a member of the *Académie des Sciences*. In 1794, he was appointed professor of the *École Polytechnique* (discussed in the next chapter). Antoine-Laurent de Lavoisier (1743–1794), also a member of the *Académie des Sciences*, was able to set the foundations of chemistry much thanks to his being a tax collector and the richest man in France. Pierre Simon Laplace (1749–1827), one of the greatest mathematical physicists of all time, studied first at the faculty of arts at the University of Caen—where he was meant to get a degree in theology, but he left without completing his studies—and moved to Paris to study mathematics under the influential Jean Le Rond d'Alembert (1717–1783), member of the *Académie des sciences* and the Berlin Academy. In Paris Laplace became a teacher at the *École Militaire*, which was founded in 1750 by Louis XV (r. 1715–1774) to train cadets of humble origin. He later became a member of the *Académie des Sciences* and of other scientific societies at which he presented many of his research results. He then was active in some of the *grandes écoles* that sprang up around the time of the French Revolution and the Napoleonic Empire, inaugurating a new era in university teaching, which we will deal with in the next chapter. Darwin, whose evolutionary theory laid the very foundation of modern biology, unwillingly studied medicine at the University of Edinburgh and then got a degree at Cambridge. After his famous discovery "Voyage of the Beagle" (1831–1836; HMS Beagle was the ship of the Royal Navy that took Darwin around the world); in 1839 he became a fellow of the Royal Society, and he carried on his research as a country gentleman. Finally, Einstein achieved revolutionary results in physics when he was an employee at the patent office in Bern between 1902 and 1909; these include an early formulation of the theory of relativity, published in a well-known 1905 article.

One could produce many more examples: they do not make a rule, but it was much more likely to find scientific innovators in a scientific society or an academy rather than in a university. Those who joined the former two institutions did not do so with the intention of obtaining a profession, a title, or social standing. Often, they already had social standing because, with very few exceptions, only noblemen could afford to practice science and belong to an academy. Those who practiced it did so out of passion, and perhaps this was one of the secrets to the success of learned societies and academies.

When conflicting intellectual or scientific social goals emerge, institutions such as scientific academies become centers of social conformism and begin declining. The *Académie royale des sciences*, according to Rhoda Rappaport, might be said to have been suffering from arteriosclerosis:

> The Academy's repeated refusals to countenance a variety of reform proposals were usually couched in terms of predictions that to change the rules would impair the quality of the institution. On a lower level, members did not hesitate to remind each other that reform might endanger seniority rights. Simultaneously, in using precedents to defend its growing autonomy, the Academy appealed to the letter of the law and the inviolability of custom. Self-interest, the weight of custom, a narrow legalism, and a sense of corporate and individual rights pervaded the membership during the decade before the Revolution. It may seem ironic that some earlier concern for merit had become a minority cause, and a radical one, even while the intellectual climate of the Old Regime was moving in the opposite direction. The Academy, however, had gone through a process not unknown in other facets of life under the Old Regime: a young institution, capable of some flexibility and some spirit of innovation, had become an established corporation.
>
> (Rappaport 1981, 253)

Outside, however, the demands for technical and technological knowledge were growing fast.

*As the academies flourished, attempts to reform the universities were
fairly unsuccessful, even though some progress was occasionally made*

To recap, the great progress made by science until the 19th century was
primarily made outside the universities.

> In the meantime the old universities continued on their way, largely
> indifferent to any except classical studies. The Cambridge University
> recognized mathematics as a major subject when it established the
> Mathematical Tripos in 1747,[7] but not for a hundred years more did
> it recognize Natural Science in a similar fashion. The mere fact that
> a university had a professor of chemistry or of experimental philos-
> ophy during this period did not necessarily mean that the professor
> was a specialist. Richard Watson, professor of chemistry at Cambridge,
> according to Miss Turner, related that at the time of his appointment
> he knew nothing of the subject; but by experimenting for several hours
> daily in his laboratory he was soon able to lecture to large and distin-
> guished audiences. He was at the same time Regius professor of divinity,
> a bishop, and rector of several parishes!
>
> (Stimson 1968, 180–181)

From the Renaissance on, the university lost its cultural supremacy and
entered a period of stagnation. Centuries would pass before it became an innovative institution and regained its monopoly of advanced knowledge. Why?

There are many reasons, starting with the religious and dogmatic character of the university. Universities remained confessional and run by clerics.
Indeed universities began to re-emerge in the 19th century in the wake of
the anticlericalism and the materialism of the Enlightenment. Another reason was their guild-like nature. As we have seen, guilds tended to protect
the social rights of their own members, rather than encourage excellence
and progress. A third, perhaps less incisive reason has to do with the social
role of the universities—a role that inevitably clashed with penchants for
cultural, scientific, and technological development. A field such as ballistics
would have greater probability of developing in an artillery school than in
a clerical seminar. All these characteristics are typical of a closed society
(in the Popperian sense).

In the last decades, historians have attempted to rehabilitate the university's contribution to science. John Gascoigne, in his article on "University"
in the *Encyclopedia of the Scientific Revolution from Copernicus to Newton*
(Applebaum 2000, 656–659), notes that many scientists who contributed to
the scientific revolution had studied, and later taught, at universities. Moreover, the university fostered those fields that later became sciences, such as
natural philosophy and mathematics. This is true: after all, the university

was the institution that offered the highest level of instruction, even if not the most advanced, and the key role of the courts and other institutions, such as the academies, that fostered the sciences was not to teach. In other words, there was little choice. Nonetheless, as we have just seen, having studied, and having taught, at a university did not mean having contributed to science. Universities till the beginning of the 19th century were teaching institutions, not research institutions, although this did not prevent their contribution to science. Mordechai Feingold (1984), for instance, relates the lively mathematical activity in Oxford and Cambridge between the middle of the 16th century to the middle of the 17th century. Feingold also points out that the study of mathematics at the universities of Oxford and Cambridge during the scientific revolution was no more traditionalistic than Gresham College in London. From the point of view of the present, it nevertheless seems amazing that so many universities in Europe contributed so little compared to scientific societies and academies; and it is not clear to what extent the activities of these universities were significant for scientific progress in that period.

Admittedly, in the 18th century the universities continued to show signs of timid progress. In Italy, for example, the University of Padua continued to excel in medicine in that century as well, thanks to collaboration with local hospitals that permitted clinical training through direct contact with patients. In Germany, too, attempts were made to reform the universities. At Halle and Göttingen efforts were made to give more momentum to philosophical studies and make them more autonomous and to encourage the exchange of ideas in learning. They paved the way for a concept of *Wissenschaft* (originally developed within the university as an objective study of the classics), according to which a university professor is also a scholar who carries out research.

Progress was also made in the teaching of the empirical sciences. For example, chemistry classes were introduced, and some scientists used demonstrative experiments to relieve the monotony of the lessons. The lectures of Georg Christoph Lichtenberg (1742–1799), physicist, naturalist, and man of letters in Göttingen who discovered the figures that bear his name in electrostatics, became legendary. At the Scottish universities of Glasgow and Edinburgh, the lectures of Joseph Black (1728–1799), one of the most eminent chemists of the century, whose discoveries include latent heat, specific heat, and carbon dioxide, were equally exciting. The University of Glasgow also supported James Watt (1736–1819), apparently the only artisan in Scotland capable of producing mathematical instruments: the smiths' guild denied him permission to produce instruments because he had not completed his seven years apprenticeship yoke [*sic*]. He repaired the university's instruments in a small laboratory, and it was there that he perfected the steam engine. Despite all, the university remained fundamentally a clerical, confessional, teaching institution of the past until after the French Revolution.

At the end of the 18th century, as Francis Bacon had predicted, the progress of science and technology was poised to bring about great social changes. In countries interested in technological progress, such as England, France, and Germany, the need arose to broaden instruction and reform the educational system. The Enlightenment ideals, and the progress in pedagogy, sparked a wide debate, in part still ongoing, involving educators, philosophers, scientists, and politicians. Which method of education could be considered the most advantageous, keeping in mind the needs of both the individual and society as a whole? What should be the specific aim of elementary, middle, and advanced schooling, and what relationship should these three levels have with each other? What should be the relationship between tradition and innovation, between pure science and applied science? Which subjects should be given precedence? In which sectors of education should the state invest funds? A number of models were proposed for the university that offered the basis for reform that would lead to its becoming that higher institution that today teaches, creates, and dominates knowledge.

CONCLUDING REMARK

Modern science developed primarily outside the university, which remained a clerical institution until the 19th century.

BIBLIOGRAPHICAL NOTES

The topics dealt with in this chapter have been treated in detail by a great number of books. Martha Ornstein's doctoral thesis dealing with the first scientific academies was a pioneering work printed privately in 1913 and published in 1928 under the title *The Rôle of Scientific Societies in the Seventeenth Century* (see p. xiii of the book). Dorothy Stimson's *Scientists and Amateurs*, published for the first time in 1948, concentrates upon the history of the Royal Society of London. Out of the many more recent works dealing with the Royal Society of London, I found Michael Hunter's *Science and Society in Restoration England* (1981) the most readable.

Ancients and Moderns by Richard Foster Jones is a classic that first appeared in 1936. It helps understand the spirit of opposition to university tradition, of Baconian empiricism and of utilitarianism, and of the Protestant reform during which the Royal Society of London took shape. Allen Debus's *Science and Education in the Seventeenth Century: The Webster–Ward Debate* (1970) complements Jones, reproducing a few works regarding a debate concerning university teaching in England in the wake of Bacon.

Hahn's *The Anatomy of a Scientific Institution* (1971) is an outstanding, thorough study of the functioning of the Paris Academy of Sciences from its foundation till Napoleon's empire. It delves in all possible contexts related to the work of its members. Taton's *Enseignement et diffusion des sciences en France au XVIIIe siècle* (1964) presents outlines of all the higher level teaching institutions in 18th-century France.

Science and Polity in France at the End of the Old Regime by C. C. Gillispie (1980) is an example of inductive history of science illustrating how lively the intellectual world was in 18th-century France outside the university, mentioning many institutions that fostered science. Despite the length of the book of more than 600 pages, and the multitude of details, it is a pleasurable read. The first chapter of Eric Ashby's *Technology and the Academics* (1963) points out, with a touch of humor, how universities, particularly Oxford and Cambridge, were so behind the times in the teaching of science and how, to begin with, this teaching was introduced in the 1800s thanks to stimuli from across the Channel.

Agassi's recent *The Very Idea of Modern Science: Francis Bacon and Robert Boyle* (2013) is an insightful book, delving critically into the philosophies of science of Bacon and Boyle in the context of the scientific revolution.

NOTES

1. Here I use the term *Academia* for university establishment.
2. http://www.constitution.org/bacon/nov_org.htm, accessed January 5, 2014.
3. The *Buccinator*, in the Roman army, blew the *buccina*—a long, circular brass tube of narrow cylindrical bore.
4. https://royalsociety.org/~/media/Royal_Society_Content/about-us/history/Charter1_English.pdf, accessed April 26, 2015.
5. I rely heavily on Ashby 1963, p. 5.
6. I rely on the articles on Cassini, Huygens, Römer, and Observatoire de Paris in *Encyclopedia of the Scientific Revolution*, ed. Applebaum (2000).
7. The University of Cambridge divides the different kinds of honors bachelor's degree by *Tripos*. The meaning of the term can be traced to the three-legged stool candidates once used to sit on when taking oral examinations. An undergraduate studying mathematics is thus said to be reading for the *Mathematical Tripos*. I rely heavily on http://en.wikipedia.org/wiki/Tripos, accessed July 23, 2014.

11 The Advent of Modern University

In the 18th century, in the wake of civil and military needs, technological schools emerged to become the basis for university reform in the 19th century

In the preceding chapters we have dealt with the religious, social, and intellectual factors that shaped, for better or for worse, universities and other institutions of higher learning. In this chapter, we will see how these factors merged into the modern university, at times enriching it, and at times producing confusion and tension.

The prime mover to the revival of the university in the early 19th century was the emerging technological higher learning. To be sure, until then, the construction of infrastructures for communications, roads, bridges, and canals was primarily the duty of the local guilds. But, as early as in the previous century, centralized France began to create state-run engineering schools, taking advantage of the progress of technology. The new technological teaching distanced itself from workshop didactics in emulation of the Renaissance, with direct contact between teacher and pupil. Here the main goal was to form experts serving the state, and a workshop does not lend itself to mass training.

In 1716, the *Corps des ingénieurs des ponts et chaussées* was founded; its purpose was to create a national network of communications. In order to train the engineers for this project, in 1747 the *École royale des ponts et chaussées* was founded; in 1775, it became the *École nationale des ponts et chaussées*, as it is still called today. It adopted a systematic, unidirectional teaching method; introduced exams, grades, and honors; and rewarded the best students. Only the most adept advanced; the only traces left by craftsmen workshops was, perhaps, a certain collaboration between teachers and students: the three best students of the course were assigned to teach their peers. For all intents and purposes, the school created a meritocracy.

In 1748, in Mézières, a town in the Champagne region, the *École royale du génie* was founded to train military engineers. Students underwent a strict selection. It was followed in 1757 by the *École des élèves du corps de l'artillerie* at La Fère, in Picardy, and in 1765 in Paris by the *École pour les élèves ingénieurs de la Marine*. In the same year the *Bergakademie* was founded in Freiberg, Saxony: it trained mining experts and promoted developments in the study of geology and in advanced technical teaching. It stimulated the foundation of the French *École des Mines* (1783), which was designed along the lines of the *École des ponts et chaussées*. All these institutions heralded the creation of polytechnic institutes for the advanced teaching of applied sciences.

At the end of the century, Enlightenment ideals, the French Revolution, and the success of technological higher learning, followed by the Industrial revolution, were among the catalysts that brought about the reestablishment of the university as an institution that monopolized knowledge. The process was nonetheless long, complex, and subject to tensions.

In 1793, the year Louis XVI was beheaded, the French National Convention reformed the country's education, including higher learning. Among other, it turned the *Jardin du Roi* into the "National Museum of Natural History" (*Muséum national d'histoire naturelle*) and the *Collège Royal* into *Collège National*; it closed all the academies, including the aristocratic *Académie Royale des Sciences* (two years later their activity was resumed under a new, all-embracing institution, the *Institut de France*), and all of the colleges and faculties, establishing a new educational system with three levels of teaching: elementary, intermediate, and higher. The Convention decreed that

> There should be established in the Republic three progressive degrees of instruction; the first for the knowledge indispensable to artisans and workmen of all kinds; the second for further knowledge necessary to those intending to embrace the other professions of society; and the third for those branches of instruction the study of which is not within the reach of all men.
>
> (Féret 1913)

The statement "knowledge indispensable to artisans and workmen of all kinds" suggests, again, a meritocratic approach. Indeed, out from the old world was dawning a new regime, and later—even more so—the Napoleonic empire, ever more in need of greater numbers of highly specialized civilian and military experts. To train them, under state control and steely discipline, were founded the *grandes écoles*. In 1794, two prominent schools were inaugurated in Paris: the *École Normale Supérieure* and the *École Centrale des Travaux Publics*, which in 1795 became the *École Polytechnique*— a source of national pride, yesterday as today.

The purpose of the *École Normale* was to train teachers. It was closed in 1795 but re-founded in 1808 by Napoleon along stricter guidelines. The term *normale* is derived from "norm," indicating the aim to form civil servants that convey "norms" (much in the Kuhnian sense), that is, teachers that form a meritocracy. In 1810, the *Scuola Normale Superiore*—a sister institution—was founded in Pisa; both still exist.

The *École Polytechnique* granted its rigorously selected pupils basic schooling in mathematics and physical science, preparing them to attend more specialized civil or military engineering schools. The selection of engineering students came to fill a gap: whereas the traditional universities came to serve the church and could recruit village young people through the village parish, engineering schools served the state. Selection, although severe, was more or less intuitive, with no particular rules of selection; aristocrats were also admitted.

To some extent, the foundation of the *grandes écoles* marked the rebirth of the university on lay basis after centuries of decline: the *École Polytechnique* granted university-level dignity to the traditionally undervalued technological higher learning. In the following two centuries, engineering increasingly gained importance in higher education. Yet, although the *grandes écoles* endeavored to be innovative, they resembled traditional universities rather than research centers, except that they dealt entirely in secular curriculum.

The *grandes écoles*, unlike traditional universities, were no longer schools for clerics. They employed illustrious scientists orphaned of their academies but with their heads still on their shoulders (Lavoisier had been guillotined in 1794 and was not the only scientist to share that fate): among them Lagrange, Laplace, Gaspard Monge (1746–1818) in the physical mathematical sciences, and Claude Louis Berthollet (1748–1822) in chemistry. Without a doubt, these scientists lent a touch of excellence. Paradoxically, whereas the goal had been to create higher level teaching institutions for the people, the *grandes écoles* were directed by innovators of aristocratic birth. No doubt the tension between excellence (as a source of innovation) and conformism (manifest in meritocracy) was present, and, inevitably, the question was posed whether the *grandes écoles* should fulfill only the basic role of specialized training, or whether they should also be innovators. Laplace wanted them to become the incubators of France's future intelligentsia. Controversy and tension reflected the complexity of the relationship between research and utilitarianism, pure science and applied science, between politics and science and, in general, among the various points of view regarding the nature and the role of science (see Belhoste 2003).

The *grandes écoles* applied military-type teaching, which put the students under pressure and in competition with each other. (This does not mean that they were militarized, even though Napoleon did militarize the *École Polytechnique* in 1804, for other reasons.) Teaching methods of this type can be effective in obtaining immediate results and selecting well-trained technicians or soldiers, but they also have the drawbacks described previously, that is, causing stress and alienating interest in the subject at hand and occasionally creating over-specialized individuals in one branch at the expense of the interdisciplinary creativity that is so vital to science (although, as Agassi 1985, 29, remarks, specialization does not necessarily require such limitation).

Pace excellency, in general, training an applied scientist often involves providing him with "normal" knowledge (always in Kuhnian terminology), that is, oriented towards solving puzzles within a preset paradigm, with the consequent danger of stagnation of the field. In today's scientific research (both "hard" and "soft"), too, merit is normally measured on the basis of predefined standards that, as Kuhn describes, can be adopted by the scientific or intellectual community following irrational factors, such as intellectual fads or social or personal needs, more aligned with conformism than to innovation. "Normal" experts, that is, meritocrats, might have been helpful in France during the Revolution and the Napoleonic empire. Creating a meritocracy, nevertheless, has its drawbacks. Meritocrats would not be very helpful or realistically appropriate, for example, in a modern Third World country that favors nuclear physicists over agronomists. The world financial and economic disaster that began in 2007 was caused above all by meritocrats, deemed "prestigious experts." Today, in addition, there are often no clear ideas as to what meritocracy should be, that is, what merits to value. For example, a democratic government, such as Italy's in the last decades, repeatedly declared that it encourages meritocracy (hoping to encourage excellence, but confounding "merit" with "excellence") without really knowing what it is and who should be favored. Much the same can be said of the "evaluation" of the quality of a university on both a national and an international level; it is conducted on the basis of arbitrary quantitative parameters that have sometimes little to do with excellence of any kind at all. Meritocracy is not necessarily compatible with the development of free thought and science. As opposed to merit, excellence lies in originality, and originality implies autonomy, the psychological basis of the ability to think outside the box, than upon conformity to whatever fashion flavors in the box. In turn, this "thinking outside the box" must be submitted to and regulated by criticism, which reduces possible harmful developments. These are the principles of controversy in an open society and should be the goal of universities.

At any rate, France's *grandes écoles* were successful on the short run and produced excellent technocrats, garnering the admiration of the entire world and inspiring the rise of technological higher learning in other countries. The United States of America, for example, followed the paradigm and founded the military academies at West Point (1802) and Annapolis (1845). Indeed, West Point was the first engineering school in the US. Even more important was the development of higher learning in Germany, where the *grandes écoles* inspired the reform of the German universities, thus creating the prototype of the modern university.

In the 19th century, the modern, state-run, and secular university
was born in Germany; its aim was to combine (academic) research
and innovation with (university) teaching

In the 18th century, the German university was in such decline that it had become a laughing stock. A number of efforts were made to reform it; the most interesting, in part stimulated by the French *grandes écoles*, took place in Prussia in the early 1800s. Nonetheless, the German model was different from that of France and represented a novelty in university teaching because it attempted, albeit rather unsuccessfully, to distance itself from utilitarianism. This was, inter alia, the outcome of debate between various contemporary philosophical currents; the many participants included Immanuel Kant (1724–1804), idealists such as Johann Fichte (1762–1814) or Friedrich Schleiermacher (1768–1834), and Romantic philosophers such as Friedrich von Schelling (1775–1854).[1] Kant's essay on *The Conflict of the Faculties (Der Streit der Fakultäten*, 1798), deals with the conflict between the traditionally "higher" (vocational) university faculties (theology, law, and medicine) and the "lower" ("liberal"—but of basic importance) philosophy faculty. More in relation to university faculties, Schleiermacher observed that nobody "would be inclined to accord a university the exclusive choice of its teachers. Universities are, one and all, so infamous for a spirit of petty intrigue that were this privilege once conceded, everybody would anticipate disadvantageous consequences" (Conant 1970, 177). Fichte, for his part, suggested that the teaching at universities should take the form of a dialogue or dialectic (Ferry et al. 1979, 172–173). The leading actor of university reform was, nevertheless, Wilhelm von Humboldt (1767–1835), a neo-humanist philosopher, man of letters, linguist, and liberal politician (he became Prussia's Minister of Public Education in 1808) who was entrusted with the task of reforming Prussia's educational system and of founding a new university in Berlin.

Humboldt's project was more or less utopian: it harked back to the neo-humanist ideal of the overall deployment of Man's individuality, rather than to the utilitarianism and to the Enlightenment's positivism. To Humboldt, the ideal university should cultivate every branch of scientific knowledge. The professor should be an exemplary scholar in his own field, as well as a good teacher; he should be erudite (*gelehrt*), and his erudition should be at the service of the growth of knowledge. He should not dedicate himself exclusively to teaching, but also to research, and his actions should be supported, but not dictated, by the state. In teaching, as in research, the professor was to have the freedom to choose his topics and methods state. For their part, the students were free to create their own curricula according to their personal interests and intellectual or professional aspirations. Lectures—the boring practice, transient ever since the invention of the printing press—were to be substituted by seminars, turning study into a creative, innovative, interesting, and pleasant conduct of debate. In other words, Humboldt wanted to extend to universities the historical role of the academies and to create a supreme institution of culture and science in which both teaching and research were conducted. To combine research and teaching is a praiseworthy and useful undertaking, given the affinity between the two activities. Both aim to increase knowledge, and this should be conducted in a rational and efficient way, above all—as Popper suggests—by asking pertinent questions. Humboldt's project, in brief, endeavored to open universities to excellence.

In 1810, the Free University of Berlin was inaugurated and still bears Humboldt's name today. The term "free" indicates that it was state-supported, with its advantages and disadvantages. One of the reasons Berlin was chosen was the possibility of taking advantage of the structures offered by the city, such as the academies of sciences and arts, the hospitals, the observatory, the botanical gardens, and the natural history collections.

The University of Berlin was paradigmatic, and to a larger or smaller degree extended to the rest of the world, creating the modern, secular university open to every field of higher learning. To be sure, the reform spirit was short-lived and dissolved with the Restoration, which rehabilitated the clergy, albeit this, too, only partially and temporarily. The development was then slow: even though, in the early 1800s, universities were becoming secular and state-run, their structure, their didactics, and many of their functions remained traditional; they often excluded the applied sciences; they rarely admitted women; they remained confessional. In Britain, despite the Roman Catholic Relief Act passed by Parliament in 1829, which removed many restrictions on Roman Catholics throughout the United Kingdom,

> No one could matriculate at Oxford without first subscribing to the Thirty-Nine Articles of the Church of England. At Cambridge, the nonconformist could matriculate but he could not hold a scholarship or a fellowship, nor could he receive a university degree.
> Consequently the "university" of London was founded in 1828 on a non-sectarian basis. That called forth much opposition on religious grounds and aroused the jealousy of the universities and the medical associations, thus delaying the grant to it of power to award degrees. A few years later King's College as its rival was established on Anglican principles.
> (Stimson 1968, 201–202)

(Note: "non-conformists" were Protestant Christians who did not "conform" to the governance and usages of the established church of England.)

The clock, in any case, could not be turned back. European society had changed profoundly; the nobility was in decline and with it the figure of the dilettante nobleman who dabbled in science. The bourgeoisie's influence was on the rise; it encouraged competition, created new needs, and took the place of the aristocracy. The need for experts, or even for literate people, grew. The academic title still offered an alternative, at least partially, to aristocratic title, and professors, as educators, were socially recognized.

After the process of unification of Germany, beginning in 1848 and ending officially in 1871, giving rise to a politically and administratively integrated nation state, the model of a state-run, secular, and positivist university spread throughout the entire country. Between 1871 and WWI, Germany was the foremost world power and also excelled in culture, science, and education. The German model was copied throughout the world.

It is impossible to follow the developments of the many different universities that emerged on the five continents; from that point on, each university developed in its own way, dictated by ideologies, philosophies, or local needs. Nonetheless, it is possible to trace the general direction of how the modern university developed, outlining its general structure, functions, and didactics. One of the basic questions is: which fields or faculty should be favored, and who should make this selection?

The modern university inherited the form of its medieval predecessor: dogmatic knowledge and an (Aristotelian) segmentation

Certain aspects of Humboldt's endeavors have been achieved, but over a long period of time and at a high cost to science and culture, resulting in a hybrid between tradition and innovation. The post-Humboldtian university maintains a quantitative concept of knowledge. And this is an Achilles heel. Moreover, the *gelehrt* academic was, and at times remains, more an encyclopedic type of scholar than an innovator. For their part, students were and remain excessively subject to professors and dependent on the textbook in vogue; they are more inclined to conform than to grow intellectually and are, above all, career-oriented.

In the 19th century scientific and humanistic fields grew or were created according to a variety of factors: economic (such as industrialization), social (such as the emergence of new social classes or of state bureaucracy), ideological, and others. The new disciplines included the "hard" sciences, such as chemistry or new branches of mathematics and physics, and "soft" sciences, such as economics, pedagogy, psychology, and sociology. As knowledge and the demand for it evolved, entire faculties could be devoted to specific fields. Moreover, new technological faculties arose, and universities of applied sciences were created. These new and indispensable institutions encouraged an exchange of knowledge and personnel between the university and industry, and this in turn stimulated research and teaching, although the universities remained basically teaching and not research institutions.

And yet all this came about in a traditional university frame. Like the earlier universities, the post-Humboldtian universities were divided into faculties, reinforced by Enlightenment encyclopedism. Despite the progress made by learning, they maintained the Aristotelian inclination to classify and fragmentize knowledge. They rarely cultivated anything interdisciplinary, and there was relatively little collaboration among the various fields, in disregard of how fields are greatly advanced by the reapplication of knowledge from one field into another. Pace Humboldt, the university teaching remained largely based on lectures. Attendance was often compulsory and lessons were and remain sometimes even more boring than medieval ones, which at least had offered a spiritual content in harmony with their era. Still today, students force themselves to write down every word the professor says, as though they were absolute truths in the authoritarianism of the positivist intellectual and the incomprehensibility of most textbooks.

A decisive aspect of the new university was the way it conceived science as a lay, but true knowledge, circumscribing about itself a new secular religion. (Indeed, Harvard's seal displays the Latin word *veritas*.) The gradual disappearance of God from science is symbolized by the famous anecdote telling that Napoleon had asked Laplace, "Newton spoke of God in his book. I have perused yours, but failed to find His name even once. How come?" Laplace is said to have answered, "I have no need of that hypothesis" (Hahn 1981, 85). God may have abandoned science, but religion and dogmatism did not. Science was taught accordingly, reaching, during the 19th century and beyond, the climax of positivism. In the first half of the century, Auguste Comte (1798–1857), a student and later an examiner at the *École Polytechnique*, formulated the philosophy of positivism, albeit in a confused manner. Like Georg Wilhelm Friedrich Hegel (1770–1831) before him, and Marx after him, Comte was a historicist, that is, he believed that history is governed by law. He claimed that human thought went through three stages of development: initially there was the "theological stage," where phenomena were explained as outcomes of divine or supernatural intervention; then came the "metaphysical stage" with explanations by means of abstract concepts; finally, humanity reached the "positive stage," that is, the age of scientific explanations. Popper, in his *The Poverty of Historicism* (1957, first published as a series of articles in 1944/45) and later in his *The Open Society and its Enemies*, heavily criticizes historicism for, among other things, one cannot know what will happen tomorrow.

Comte's positivist empiricism marks the zenith of scientific dogmatism, to the point that he envisaged a new atheist religion of humanity, with scientists as its priests and himself as the high priest. (He was repeatedly hospitalized for mental illness.) The medieval professor, almost always a priest, was thus substituted by the positivist professor, a secular priest of science. With a bit of imagination the process is reminiscent of the replacement of priests by rabbis in ancient Judaism. Even if Comte's philosophy was not initially admitted at the university due to its anticlericalism, it paradoxically indicated how clerically science remained perceived. As Polymath Michael

Polanyi (1891–1976) puts it in his *Science, Faith, and Society* (1946), in the modern university

> No one can become a scientist unless he presumes that the scientific doctrine and method are fundamentally sound and that their ultimate premises can be unquestioningly accepted. We have here an instance of the process described epigrammatically by the Christian Church Fathers in the words *fides quaerens intellectum* faith in search of understanding.[2]

Polanyi may have gone too far in suggesting that a scientist must believe that the scientific doctrine and method he embraces are fundamentally sound and that their ultimate premises have to be unquestioningly accepted. Yet he is right in comparing the university to a church. Popper's pupil, Alan Musgrave adds,

> In religion, the dogmas of articles of faith are usually embodied in a sacred text or 'Bible'. So also in science: Newton's *Principia* or Lavoisier's *Elements of Chemistry* are early examples of scientific Bibles—nowadays we have the up-to-date textbook.
> [. . .]
> Religions have their initiation rites, when the student is recognized as a mature believer. So too in science, where the most important initiation rite is the award of the PhD. Religions have their religious festivals, where the faithful gather together to reinforce their collective commitments. Science too has its religious festivals: they are called 'conferences'.
> Religions have systems for ensuring that nothing that might endanger the faith can get publicity. Science also has censorship systems, above all the device of refereeing papers submitted for publication.
> [. . .]
> Religions have their high priests, who have special responsibility and authority for interpreting and transmitting the faith.
> (Musgrave 2009, 51–52)

Musgrave acknowledges this regrettable state of science, but, beyond Polanyi and Kuhn, he endeavors improvement: "Religions are occasionally beset by heretics who question the articles of faith—these are strongly discouraged and if they persist, excommunicated. Science too has its heretics: attempts are made to suppress them, as if this fails, they are branded cranks and ignored" (Musgrave 2009, 52). Galileo is the best known heretic in the history of science. In a period in which one did not make clear distinction between religion and science, he was condemned for insisting on such a distinction. Today, many outstanding scholars are barred from universities, and even from publishing by high-status publishers of "prestigious" scientific reviews or publications, for their non-conformism.

The university confuses between vocational and liberal learning

Despite Humboldt's hopes and efforts, the 19th-century university remained primarily a teaching institution with vocational endeavors, despite that most fields, old and new, were hardly vocational. Yet, in a society still to a great extent illiterate, higher learning in any field helped employment prospects. Often, philosophers or physicists, like many other graduates, were unable to pursue their "profession," as opposed to lawyers, medical doctors, or engineers, but, thanks to their education, could get a job in state administrations or in schools. In the last case, to be sure, the true profession was that of a teacher, and universities even established and developed specific pedagogy curricula.

The confusion between liberal education and vocational learning was clearly expressed in the introductory lecture held by the German physician, physiologist, physicist, and mathematician Hermann Ludwig von Helmholtz (1821–1894) at the University of Heidelberg in 1862: "Whoever, in the pursuit of science, seeks after immediate practical utility, may generally rest assured that he will seek in vain [. . .] Each individual student must be content to find his reward in rejoicing over new discoveries" (Ashby 1963, 25).

The tension was, of course, less felt in the new emerging technological higher learning institutions. The teaching of applied science emerged, as seen, on the continent: Germany followed France and, as early as in the first half of the century, *Technische Hochschulen* (abbreviated "*TH*"s) with the dignity of universities were founded in capital cities of German states, indicating the bureaucracies' demand for technically trained personnel. In Switzerland the ETH-Zurich (*Eidgenössische Technische Hochschule Zürich*) was founded in 1854, and in the US, MIT (Massachusetts Institute of Technology) came to being in 1861. They were, at least initially, vocational.

"Though the *TH*s hardly posed a direct threat to the traditional universities, on account of their differing approach to science, many university professors feared and resented them" (McClelland 1980, 237). And in the 19th century, European universities were not, on the whole, the appropriate institutions for applied science. England remained particularly behind as far as technological higher learning is concerned:

> In nineteenth-century England it was the artisan and lower middle-class manufacturers who first took to science; the mechanics' institutes served the one, the University of London and the colleges which had sprung up in Manchester, Birmingham, Leeds, and elsewhere served the other. [. . .] And since Oxford and Cambridge were the preserve of gentlemen, here was a suitable excuse for continuing to neglect science there, even despite the discoveries of Dalton and Faraday. [. . .]

A great deal of credit for implanting the spirit of scientific research into British higher education goes to the Prince Consort. Undeterred by the indifference of English academics, he used his influence wherever he could to support science. One of his contributions was to help to establish in London the Royal College of Chemistry in 1845. The College arose in this way. Liebig's researches in Giessen on artificial fertilisers had caught the imagination of some wealthy English landowners and (following a visit from Liebig in 1842) the Royal College was set up on the model of Giessen, under an imported German, A. W. von Hofmann.

(Ashby 1963, 31–32)

The Royal College of Chemistry finally merged into the Imperial College of Science and Technology, created as late as 1907.

Before the 1870's there was neither an adequate supply of pupils trained in science from schools nor an adequate demand from industrialists for graduates. Training in technology was through apprenticeship, on the job; and any formal training in colleges was regarded with suspicion, as likely to lead to the disclosure of 'know how' and trade secrets.

(Ashby 1963, 56–57)

Nonetheless, British higher learning institutions were still very much ruled by a medieval, "closed" way of thinking. Oxford and Cambridge remained primarily incubators for the managerial class, although, over time, the differences between them and other higher learning institutions faded.

The university's importance increased through the introduction of research, but the teaching methods remained traditional and of medieval origin

In general, 19th-century universities, as many universities today, were not, and often are still not, equipped to conduct research. The students who wanted to pursue a specialization became subject to the tenured professor, running the risk of servility and pressures to intellectual dishonesty. Universities remained basically teaching institutions throughout the 19th century, although a number of them began, however timidly, to conduct research, fulfilling Bacon's and Humboldt's dream and increasing the dignity of the university, even though their role as centers of research and innovation was more a product of the 20th century. There were, of course, exceptions: France's *grandes écoles*, for example, thanks to the outstanding scientists who taught there, managed, initially at least, to do research and produce innovation thanks to scientists such as Laplace or Jean-Baptiste Joseph Fourier

(1768–1830). The latter formulated the mathematical transform that took his name and found a variety of applications in physics and engineering.

Moreover, post-Humboldtian research increasingly took hold, particularly toward the end of the 19th century. Professors, encouraged by the Humboldtian spirit and by social developments, developed their field and endeavored recognition. Even though universities had still not officially established research in the 19th century, they did occasionally encourage it and contributed more to innovation than they had in previous centuries, in Germany in particular. A number of other famous examples follow: In the 1820s, Justus von Liebig (1803–1873), who had studied in Paris under physicist and chemist Joseph Louis Gay-Lussac (1778–1850), set up at the University of Giessen a chemical laboratory that had close ties with teaching and formed the nucleus of a school of chemistry. In 1872, following the developments of non-Euclidean geometry, the young mathematician Felix Klein (1849–1925) held a famous opening address at the University of Erlangen, which became the basis of the "Erlangen Program," to classify the various geometries. The program was developed in various universities, and the University of Göttingen, where Klein later transferred, became one of the world's leading most centers of mathematical research. In 1872, the building of Oxford's Physics Clarendon laboratory was completed, and two years later, the Cambridge Cavendish Laboratory dedicated to research in the physical sciences was inaugurated. In 1879, at the Leipzig University, the physiologist and psychologist Wilhelm Wundt (1832–1920) founded an institute of experimental psychology with a research laboratory. One of his students was young Durkheim, who, after returning to France, applied the same empirical method to social science. Durkheim taught the very first course in sociology when he was professor of pedagogy at the University of Bordeaux. In 1896, to repeat, the latter university established the first chair of sociology, followed, in 1913, by Paris. Durkheim constantly performed research, and his top achievement was his work on *Suicide* (1897)— an empirical study and classification of the phenomenon. There are many more examples: the university was on its way to monopolizing science, and the more it adopted research, the more cosmopolitan it became. And, in a famous lecture held in 1918 in Munich, "Science as a Vocation," Max Weber already took it (wrongly) for granted that only professors can do research and these are and will remain few.

Many innovations came, however, also from beyond academia, above all from industry and the military. Governments recognized the importance of research and the need to encourage it, but they often did so outside the university system. For example, 1911 saw the foundation in Germany of the *Kaiser-Wilhelm-Gesellschaft zur Förderung der Wissenschaften (KWG)*— literally, the "Kaiser Wilhelm Society for the Advancement of Science," created to work alongside Prussia's Royal Academy of the Sciences: it came to be composed of research institutes in specific fields. A large amount of financing was provided by patrons, and the *KWG* was a self-regulated

182 The Advent of Modern University

institution under state control. The goal was to advance the natural sciences, supporting famous scientists to freely conduct research. In 1917 the Berlin *KWG institute* for physics was founded, under the direction of Einstein. In 1933, after the Nazis came to power, Einstein was forced to abandon Germany and was welcomed by the newly founded *Institute for Advanced Study* in Princeton, New Jersey. After the end of WWII, the *KWG* was renamed the *Max-Planck-Gesellschaft (MPG)*, adopting the name of one of the most famous physicists of modern times, Max Karl Ernst Ludwig Planck (1858–1947), originator of the quantum theory, according to which electromagnetic energy could be emitted only in quantized form. Planck also gave his name to one of the basic constants in physics: "h"—the ratio between energy and frequency of an electromagnetic wave. Today, the *Max-Planck-Gesellschaft* supports every field of the natural, human, and social sciences.

The 20th-century university is meritocratic, and so it remains today

It is indeed in the US that the Humboldtian spirit managed to flourish perhaps more than in Europe. Two Paladins were Daniel Coit Gilman (1831–1908), first president of the Johns Hopkins University, and Charles William Eliot (1834–1926), president of Harvard.

> At the time Eliot had become president of Harvard in 1869, relatively few additions had been made to the collegiate fare that was offered at Harvard and other colleges before the Civil War. Based on the classical tradition and with a heavy emphasis on developing a reading and writing knowledge of Latin and Greek, it was a closely circumscribed program. The introduction of courses in modern foreign languages, history and natural science was just starting (Eliot was himself a chemist).
>
> By introducing what was called "the free elective system," whereby all that was necessary to obtain a degree was a passing grade in sixteen courses. Eliot laid the base for the growth of instruction in a variety of fields. The proliferation of college courses started under his regime at Harvard, and was duplicated in many other colleges as the elective system grew in popularity.
>
> (Conant 1970, 20–21)

The introduction of the elective system opened the way to research and, consequently, to specialization. Yet attempts to combine liberal and vocationally oriented studies seemed to be too utopian, and the elective system was dismantled by Eliot's successor Abbott Lawrence Lowell (1856–1943), who presided over Harvard from 1909 to 1933 and introduced frameworks that directed the programs of studies.

Lowell replaced the elective system by concentration and distribution requirements, accompanied by a tutorial system. Yet limiting studies must follow some kind of philosophical or metaphysical factors, explicit or implicit. Harvard became less liberal and more meritocratic, and the system extended to universities in the rest of the country. Moreover, "since the 1890s there had developed in the larger universities the tradition of employing graduate students as half-time teachers in elementary courses. Abolishing this practice would unquestionably improve the freshman course, but at the price of a drastic diminution in the number who were working for a Ph.D." (Conant 1970, 66).

In the second half of the 19th century, American taxpayer's money had started to flow to universities to support applied research. The physical chemist Gilbert Newton Lewis (1875–1946), Dean of the college of Chemistry at Berkeley,

> had no sympathy with this trend. He said that it was unnecessary and unwise to attempt to persuade a state legislature to support research because of its practical results. There was no guarantee that the ills of the farms would be cured by the work of the professors; indeed, from time to time they might be made worse. The propaganda emphasizing applied research was likely to boomerang. At all events, the University of California would have none of it. Lewis and his friends on the faculty were going to demonstrate that the way to obtain public support was to appeal to the pride of the citizens. They would make the university so famous in the world over that the voters of California would be glad to see public funds expended in pure research.
>
> (Conant 1970, 66).

Research in the US before the Manhattan Project is described in American novels as a mixture of private, semi-public, industrial, government, and fringe-academic, and the background detail of fiction is normally fairly realistic and seems sociologically true, fictitious characters notwithstanding. Thomas Edison (1847–1931)—who, incidentally, did not study at a university and, moreover, attended school for no longer than three months—performed his research in a combined workshop and laboratory at Menlo Park and later in Orange, both in New Jersey. Writer and philosopher Paul Carus (1852–1919) is an example of an intellectual who tried to support research privately. And then there were the well-heeled foundations, along with the US geological survey and as such.

Nonetheless, research remained essentially an extra-university activity until after the end of WWII. Impetus was given by the war and by the scientific research it necessitated, as well as by the "GI Bill," formally called the "Servicemen's Readjustment Act," which was signed into law in 1944 by President Franklin Delano Roosevelt (1882–1945, in office from 1933 till his death). The aim of the GI Bill was to facilitate the re-entry into civilian

life of US veterans returning from the war in Europe and the Pacific, supporting, inter alia, their studies. This led to the expansion and increase in a number of American universities, which were also offered government financing
under certain conditions. To obtain this financing, they had to demonstrate
their level of competence, and they did so by introducing research. They
increasingly taught theoretical subjects that were considered more "scientific" and demanded that their professors publish papers in scientific journals. Thus, after the war, the university expanded and adapted itself to the
new demands; this model of the university for the masses extended throughout the world. And it is in this sense that it confirmed itself as the universal
home of knowledge.

Science evolved, but the university remained basically a closed society

Some of the previous quotations are taken from the autobiography of James
Bryant Conant (1893–1978), chemist and Lowell's successor as president
of Harvard from 1933 to 1953, and a central figure in 20th-century higher
education. Conant, as one learns even from the title of the book, *My Several Lives: Memoirs of a Social Inventor*, must have been a most ambitious
man, and he encouraged, from the very beginning of his office at Harvard,
the formation of a meritocracy: "The standards for promotion were not
high enough to suit me" (Conant 1970, 82). But, what were, for him, the
standards of promotion? "If the permanent professors were the most distinguished in the world, then the university would be the best university"
(Conant 1970, 83). "Distinguished" here means approval and success as
appraised by common sense and arbitrary normative standards.

One contribution of Conant to US and world education was to the establishment, in 1947, of the Educational Testing Service (ETS), today's larg-
est, private, nonprofit educational testing service that develops standardized
multiple choice exams for US schools and higher education as well as international tests, such as the TOEFL (Test of English as a Foreign Language).
The ETS homepage relates that

> In creating ETS, our founders brought to life a concept proposed a
> decade earlier by Harvard University President James Conant. Conant's
> belief was that a single organization devoted to educational research
> and assessment could make fundamental contributions to the progress
> of education in the United States.[3]

"The establishment of ETS," Conant declares (1970, 432), "was part
of an educational revolution in which I am proud to have played a part."
According to Conant (1970, 424), "The aptitude, not the schooling was to

be counted," where "aptitude" is "inherent capacity," which, as it is today and generally accepted, at the end is the capacity to develop competence. Despite his idealism and good will, Conant contributed to making university more "normal," in the Kuhnian sense, and more vocational, and thus to perpetuate a central historical tension described in the present book. As a result,

> A particular degree is usually intended to fulfill two functions. One is to enrich the knowledge and enhance the skills of those enrolled. The other is to eliminate those who do not have the kind of talent which the leaders of a profession or vocation believe to be essential for success. The second objective is rarely set forth as crudely as I have phrased it. Rather, it is customary to speak of selecting those who would be worthy members of a profession. [. . .] The diploma is thus, in fact, a certificate of vocational competence.
>
> (Conant 1970, 632)

The state of the present university with all its tensions could not have been expressed better: vocational, meritocratic, and corporative. Catholic Historian Christopher Dawson (1889–1970), in his *The Crisis of Western Education* (2010 [1961]) even suggested that higher education should return to its religious roots. The result, in any case, is well described by Popper's remark quoted earlier in relation to Plato's educational system as an obstacle course:

> Instead of encouraging the student to devote himself to his studies for the sake of studying, instead of encouraging in him a real love for his subject and for inquiry, he is encouraged to study for the sake of his personal career; he is led to acquire only such knowledge as is serviceable in getting him over the hurdles which he must clear for the sake of his advancement.
>
> (OS 1, 135)

The difference is that today academia monopolizes science, and therefore has an enormous responsibility with no rival institutions as there were during the Renaissance. Academia can no longer allow itself to remain a "closed society."

The solution to the problem is to open the university and to encourage criticism. This cannot be done with a general reform because every country has its own peculiar university system, and every university is "closed" or "open" to a lesser or greater degree. Every university should have clear ideas concerning its scope, whether vocational or liberal, in order to afford students transparency. Bearing all this in mind, every responsible university might reform itself so as to encourage the autonomy of its members and a more responsible attitude towards science and society.

BIBLIOGRAPHICAL NOTES

Ferry et al.'s *Philosophies de l'université. Textes de Schelling, Fichte, Schlei-ermacher, Humboldt, Hegel* (1979) is a valuable translation and collection of writings and debates around the reform of higher learning in Germany.

Belhoste's *La formation d'une technocratie: L'École polytechnique et ses élèves de la Révolution au Second Empire* (2003) concerns the development of the *École* in the 19th century, including the initial debates around its role. The text is as intricate and complex as the subject itself.

Conant's 700-page autobiography, *My Several Lives. Memoirs of a Social Inventor* (1970), very well written, is, despite criticism, a most valuable document concerning the development of American higher learning in the 20th century.

Suzanne Mettler's *Soldiers to Citizens* (2005) is a pioneering work investigating the effects of the "GI" bill, concentrating nevertheless more on the sociological and personal aspects, rather than upon the developments of US universities.

NOTES

1. Compiled and translated into French by Ferry et al. 1979.
2. http://archive.org/stream/sciencefaithands032129mbp/sciencefaithands 032129mbp_djvu.txt, accessed July 27, 2014.
3. https://www.ets.org/about/who/, accessed September 1, 2014.

Appendix 1
Galileo and the Medici
Post-Renaissance Patronage or
Post-Modern Historiography?

In 1608, the telescope was invented in the Netherlands. Galileo, at that time mathematics professor at the University of Padua, in the Republic of Venice, heard about the discovery and constructed his own instrument. He trained it to the sky and made a series of startling discoveries that inaugurated modern astronomy. He then took advantage of these discoveries to approach the Medici court in Florence and gain their patronage. A new phase began in Galileo's career, and he was able to return to Florence as a courtier taking advantage of the Medici's patronage.

At the beginning of the eighties of the last century the issue of "patronage" began to arouse scholarly interest and gained importance. Galileo became a test case: his importance, and the importance of patronage—that of the Medici in particular—go beyond the historical junction of the scientific revolution and have corollaries in the more general attitude to science and knowledge. A leading historian of science, the late Richard Westfall (1924–1996), published in 1985 in *Isis* an interesting article under the title "Galileo and the Telescope" (Westfall 1985) arguing that Galileo's main concern then was less astronomy than the telescope's capacity to ensure his own future at the Tuscan court.

Westfall lamented that, in general, history of science had been excessively dominated by 19th-century concepts. He suggested drawing more on 17th-century concepts whereby "the subtle alchemy of patronage transmuted an object of science into an *objet d'art* to amuse and flatter a prince" (Westfall 1985, 15). He concluded that patronage could well have been the most pervasive institution of pre-industrial society as well as an avenue leading us into the fruitful social history of the Scientific Revolution.

Westfall's claims are interesting and—although doubtful—worth consideration. His challenge was welcome, and in the following years, Galileo's ascent to the Tuscan court has been meticulously investigated. Historians and sociologists of science have tried to reconstruct, to use the current jargon, the "strategies" involved, such as "microphysics of patronage" and "self-fashioning of a client versus his patron," flavored by a variety of "practices" drawn from realms such as etiquette, rhetoric, art, mythology, and emblems. A substantial contribution has been given by Mario Biagioli (1993), who maintains that science was part and parcel of his career and self-fashioning at the Tuscan court. In general, the standard trend in the history of science claimed that science would not have evolved the way it has had it not been for patronage and even that in Galileo's case, and in Early Modern Europe in general, patronage and science were more or less the same.

Two difficulties make me question these views. The first is historical: In the Galilean case, patronage ultimately played a relatively restricted role, if any, in advancing science. The second difficulty is philosophical: patronage—in particular that offered by a court—entails conformism. How can this be reconciled with science's demand for innovation?

One is tempted to explain the first difficulty by the second, but the historical context is much more complex. Let us look at some of its aspects.

Patronage and Science

Galileo's move to the Tuscan Court was, at the beginning, advantageous to both sides, to the Medici in particular. In a period of utter decline, the presence of Galileo at court enhanced their prestige: they could present themselves as patrons of the new, emerging science and pursue the traditional cultural policy of the dynasty. But this was no more than a temporary aura, and, in the long run, Galileo's presence at court did not produce any particular advantage either to the Medici or to Tuscan culture.

Galileo's endeavors were mainly financial, namely, to be granted enough leisure to complete his scientific work without having to teach. His salary, incidentally, was not paid by the court but by the University of Pisa—a fact that raises a question concerning the extent and nature of the Medici patronage (OG 19, 233–264). Independently of the source of money, Galileo got all he asked for and more: in addition to good financial treatment and no commitment to teach, he was able to take advantage of all the services a court could offer. Furthermore, he was totally free to proceed with his work, an exceptional situation, as we shall soon see, in Tuscan Post-Renaissance patronage.

The outcome, however, was disastrous. In 1616 the Catholic church prohibited Galileo from teaching the Copernican theory, and in 1633 the Roman Inquisition sentenced him to life imprisonment. The Grand Duke of Tuscany was only able to offer him his carriage to go to Rome and to put at his disposal the services of his embassy. History cannot be based on "ifs," but all this would probably not have happened had Galileo remained a well-paid civil servant of the relatively strong and independent Republic of Venice. Moreover, much of Galileo's contribution to science was made prior to his return to Tuscany: his major work, the *Dialogue* of 1632, is essentially a popular presentation, on the highest literary level, of previous thoughts. Even without Galileo's campaign the heliocentric theory would have established itself, thanks to the contributions of astronomers such as

Kepler, Huygens, or Newton—and in a less traumatic way. Taking Galileo as a test case, then, scales down the importance of patronage.

Why, then, give so much importance to patronage? To answer the question one should consider a broader historical and historiographical context. As far as history is concerned, a look at the development of the Medici patronage could be helpful, and this requires extending the discussion from history of science to history of art, more specifically, to history of art patronage.

The basic question is,

What was the purpose of the Medici patronage?

Let me make one remark before I attempt to answer: The very concept of patronage is both broad and ambiguous. According to the Collins English Language Dictionary, it is generally defined as "help and financial support given by someone to a person or group"[1] to enhance enterprises such as science, art, or culture. But consider, for instance, a later case—that of the pioneer of bioelectricity, Luigi Galvani (1737–1798). After resigning his chair at the university of Bologna, to avoid taking an oath of loyalty to the Napoleonic Cisalpine Republic, Galvani was sheltered by his brother. Can we call this patronage? And can one call the salary paid grudgingly by the University of Pisa "court patronage"? The issue gets even more complex if one considers that the Medici patronage began in the 15th century, lasted three centuries, and involved at first the arts and letters and only later the sciences. Although the literature describing it is enormous, it is interesting that among all the works concerning the Medici patronage that I was able to see, none answers clearly or even raises the basic question of its purpose. I am not a Medici expert, nor an historian of art, but it seems to me obvious that their style of patronage varied at different times and with different rulers. It began as a successful private enterprise and developed into a less successful state project.

The first Medici to rule Florence were bankers whose motivation was protecting their finances. The complex structure of the Florentine *Comune*, which some historians call a "League of Mafia families," needed—inter alia—to be on good terms with artists and their guild (Trexler 1980, 27). The first famous Medici "godfather," in the first half of the 15th century, was Cosimo the Elder who promoted the arts in an enlightened way. He gave his clients total freedom, at least occasionally, even if this meant the deliberate ignoring of current morality. An example of this is Donatello's "David," commissioned by him in 1434, the first life-size nude to be cast in bronze since Classical times, and with a playful, sensuous, and androgynous body. Cosimo's patronage, in addition to his personal taste and to local political calculations, seems to have sprung from religious motives. His earthly enterprises not always having been spotless, he may have hoped to redress the balance with pious deeds. One of his major sponsorships was the restoration of the Dominican Cloister of San Marco in Florence. In its cells one can still admire Beato Angelico's (c. 1395–1455) wonderful, meditative frescos, and Cosimo's own cell testifies to his spiritual concern.

Cosimo's heritage reappeared in his legendary grandson, Lorenzo "the Magnificent," who supported artists and men of letters in the state's interest in addition to his own personal one. On the personal level, Lorenzo was a learned man who collected rarities, sponsored crafts neglected by traditional patrons, and enriched the Medicean libraries with rare manuscripts. On the political level Lorenzo was more ambitious than his grandfather and endeavored to win for Florence and its scattered territories the cultural leadership of Italy. He used art and artists for diplomatic and propaganda purposes and strengthened ties with other princes and states by offering artistic advice and art objects and by recommending artists.

Yet the outstanding historian André Chastel (1912–1990) argued, under the provocative title "*Le mythe de la Renaissance: age d'or et catastrophes*" ("The Renaissance Myth: Age of Gold and Catastrophes," a chapter in Chastel 1959, 341–351), that Lorenzo's time ran far less smoothly than his legend suggests: not only was the quality of intellectual production lower than that of the earlier Renaissance, but Lorenzo's cultural enterprise seems at times to have been excessive and to have been a burden on the family's and state's finances. Together with wars and plagues, it brought social unrest, which—after Lorenzo's death in 1492—forced the Medici into temporary exile and brought Girolamo Savonarola (1452–1498) to power.

It is interesting to note that Savonarola came from the very San Marco Cloister that had been restored by Cosimo the Elder, and that Savonarola's Dominican followers in San Marco gave birth to a reactionary, anti-humanistic, and later anti-Copernican trend directly related to Galileo's trial: one of the main anti-Copernican figures in this trend was the Dominican Giovanmaria Tolosani (c. 1470–1549; see Camporeale 1986). The first serious challenge to Galileo and Copernicanism was launched in 1614 from San Marco. I mention this as an example of how unpredictable the results of patronage can be. The Florentine decline at the end of the 15th century, by the way, is depicted in Sandro Botticelli's (1445–1510) later works and confirmed by the fact that leading Tuscan high Renaissance artists, including Leonardo da Vinci and Michelangelo, had to try their luck elsewhere. A century later much the same happened to Galileo, when, as a young, innovative mathematician and academic, he had to leave his homeland for Padua.

It seems therefore clear that at least part of the aura surrounding Lorenzo's patronage was artificially produced, by him or by later generations. Furthermore, during the period in which Italian rulers were competing with each other to raise their prestige and embellish their cities and palaces, European nations were taking shape. Europe and the World were going through epochal changes and entering the modern era. Fine arts could contribute little to assist the Italian principalities to keep up with these developments. The Florentine Niccolò Machiavelli (1469–1527), at the beginning of the 16th century, foresaw all this and suggested political remedies. Chapter 24 of his celebrated *The Prince* (*Il Principe*, written in 1513 and published for the first time in 1532) is "an exhortation to liberate Italy from the barbarians," expressing the wish that Italy becomes a great European nation.[2] When the Medici returned from exile to Florence, Machiavelli offered his good offices to the Medici, but he was ill-treated and shunted aside.

That is the picture of Tuscany at the beginning of the 16th century. During this post-Renaissance century the Medici became the absolute rulers of most of the Tuscan territory, received the titles "Duke" and later "Grand Duke," and hoped, at times pathetically, to obtain a royal crown. They deluded themselves that culture could be instrumental in achieving this ambition, and this opened an interesting new chapter in their patronage: post-Renaissance Patronage.

The main aim of the first Tuscan Duke, Cosimo I, was to emulate the great European powers, Spain in particular (see Forster 1971 and Segre 1991a, 7–9 and 144–145). Cosimo I, unlike Cosimo the Elder and Lorenzo the Magnificent, was not an intellectual. His goal was not so much to encourage culture as to use it for Tuscany's and his own glorification. He introduced a complex art of patronage that was carried on by his heirs and lasted over a century.

Briefly, Duke Cosimo I raised culture to the status of a major official state project. This included financing, and at the same time putting Tuscan cultural institutions (universities and academies in particular) under strict state control so that they could be used for political purposes, and also to avoid possible opposition. He invested in projects that could glorify the dynasty, such as opening botanical gardens, which fitted well into the absolutist extravagance of his day, or establishing new university chairs that would attract leading scholars. Yet he did not succeed, for instance, in bringing Andreas Vesalius, the leading anatomist and physician to Charles V, to the University of Pisa.

Most importantly, as far as historiography is concerned, Cosimo I allowed an artistic genius, Giorgio Vasari, to supervise the state's artistic interests. One of Vasari's undertakings at court was to glorify Tuscan culture and art under the patronage of the Medici. His *Lives of the Artists*, considered to be the beginning of modern history of art, was soon criticised for paying too much attention to Tuscan art and neglecting art produced elsewhere on the peninsula. Bolognese historian of art Carlo Cesare Malvasia (1616–1693), for example, in his *Le pitture di Bologna* (1686) criticized Vasari for, among other things, neglecting non-Tuscan artists, particularly from Bologna (Malvasia 1969, 1–2). Vasari used his literary ability to inflate and spread the myth of the Medici as enlightened patrons (see Forster 1971 and Rubin 1995, 197–208).

Vasari, by the way, played a part—without knowing it—in the creation of the Galileo myth (although he died in 1574 when Galileo was ten years old). His hagiographical style of writing was adopted by Galileo's influential follower and earliest biographer, Vincenzo Viviani, in his *Vita* of Galileo written in 1654 (*OG* 19, 597–632). Viviani was strongly influenced by Vasari. One sentence of his biography could almost have been copied from Vasari's *Life of Michelangelo* (see Segre 1989 and 1991a, chap. 7).

The image of Tuscan culture created by Vasari and the Medici post-Renaissance patronage in general may have been impressive but utterly failed to achieve their political aims. The country was sinking into irreversible political, artistic, and cultural decadence. The strict control under which intellectuals and artists were forced to work was unpopular and harmful.

This, then, was the state of affairs when Galileo joined the Tuscan court in 1610—much tradition, some decadent splendor, and very little substance. One can well understand why Grand Duke Cosimo II was very happy to have Galileo at court, and why, as an exception to traditional post-Renaissance Medici patronage, he imposed no restrictions on him. Yet, as I pointed out previously, even this did not help.

Incidentally, Cosimo's heirs, in the middle of the 17th century, tried to return to the traditional post-Renaissance policy of control, censoring the work of the Galilean followers they supported (Segre 1991b). But this, too, as one of the last prominent members of the dynasty, Prince Leopold de' Medici, frankly admitted, did not bring the desired results (Middleton 1971, 316).

In view of all this, to claim that patronage and science were, at that time, more or less the same thing requires considerable imagination, or perhaps faith in somewhat speculative theories in sociology and anthropology, in addition to taking for granted past historical descriptions that have been shown to be more legendary than true. How was this possible?

From post-Renaissance patronage or post-modern historiography

In the 19th century, as modern history of science was establishing itself, Galileo was largely depicted as the mythical founder of experimental science, and the only social aspect of his work considered was his trial, which suited the anticlerical feelings of the day (Segre 1998a, 393–396). Yet, as the field developed, historians began wondering what criteria to adopt when choosing among historical facts. Also, in the same century, the founder of positivism, Auguste Comte, invented the term *sociology*, and Emile Durkheim established this field, conjecturing that "social facts" (to use a Durkheimian term) are the basis for human actions. Among the countless historiographies suggested, the presentation of science as a social occurrence began gaining ground. Thomas Kuhn was perhaps most instrumental in establishing it in the second half of the 20th century. Kuhn drew attention to the relevance of the social aspects of science and particularly to its professionalization (I find it hard, nevertheless, to consider mathematicians such as Copernicus, Galileo, or Newton as scientific professionals).

Historians of science became increasingly interested in the social developments related to science and, instead of chronological facts and discoveries, began speaking of "practices." Substituting practices for facts moved the discussion to the sociological domain, but the difficulties remained. The hoary issue of experimenting, for instance, re-emerged, with the difference that instead of speaking of experiments, one spoke of "practice of experiments" and concentrated more on the experimenters that on the experiment itself.

The myth of Galileo as the founder of experimental science, like Viviani's hagiography of Galileo, or Vasari's myth of the artistic patronage of the Medici, all share one problem that post-modern terminology calls "conflict between science and narrative." This is the starting point for the French philosopher Jean-François Lyotard, who, in 1979, published his influential book *The Postmodern Condition*, in which he proposes a postmodern approach.

Postmodern is an even more vague concept than patronage. The term was invented in the 19th century in relation to art and was adopted in different fields with different meanings. It distances itself from what it considers the monolithic approach of modernity, no longer pertinent in a culturally diverse and fragmented world such as ours. It has produced interesting results, such as *Learning from Las Vegas* by Robert Venturi et al. (1977), where, in a study of Las Vegas's architecture, the authors show how a mixture of styles can be attractive.

In philosophy, Lyotard and others express disillusion with the Enlightenment's rationality and reject absolute standards and truth: knowledge can only be relative. They favor narrative without meta-narrative (i.e., narrative without the classic theory of rationality). They consider science as strictly human and, therefore, suffering from human bias.

Westfall's article was well accepted in the new trend, despite, or perhaps thanks to, the ambiguity of his claims. Assuming, for instance, that Galileo thought first of his career, as Westfall claims, this does not mean that patronage was as important, or as pervasive as Westfall claims. Post-modern historiography can, however, digest, and even welcomes, irrational and confused discourse. Westfall's article received a prize from the History of Science Society and inspired many other works emphasizing the importance of the Medici Patronage, contrary evidence notwithstanding.[3] Incidentally, quite a few articles that in the following years were awarded prizes by the History of Science Society were clearly post-modern.[4]

Conformism, then, seems to be the common denominator between post-Renaissance patronage and post-modern historiography, and that is why a postmodern historian would exalt post-Renaissance patronage with a clear conscience. I, myself, find it difficult to agree and find it still helpful to apply the old, modest, Popperian approach. Had historians posed the question, for example, what was the purpose of the Medici patronage? or other related, specific questions, we would have found ourselves in a much better position to give a balanced judgement. And this would probably indicate that patronage had more incidental consequences than have been presented in the past decades.

NOTES

1. *Collins Cobuild English Language Dictionary* (London: HarperCollins Publishers, 1987).
2. http://www.constitution.org/mac/prince00.htm, accessed September 11, 2013.
3. The article was awarded the 1997 Zeitlin-Verbrugge Prize from the History of Science Society.
4. An example of a postmodern formulation is Paula Findlen's "Controlling the Experiment: Rhetoric, Court Patronage and Experimental Method of Francesco Redi" (1993). The author suggests (p. 41) labeling Francesco Redi—a naturalist working at the Medici court in the middle of the 17th century—rather than a leading contemporary biologist, "a courtier who deployed the natural and human resources that his environment offered to shape experimental narratives that met the expectations of a patrician and largely court-based audience." Having myself studied Redi's works and manuscripts, I still feel that he was a leading biologist, and Findlen's description is an example of trendy, postmodern jargon.

Appendix 2
Kuhn, Meritocracy, and Excellence

Two terms are frequently mentioned in relation to science and education: *meritocracy* and *excellence*. They are often confused, although they express conflicting concepts—meritocracy normally aims at conforming to a framework, whereas excellence breaks away from it. I argue that Kuhn realistically describes science as a structure that seeks merit, rather than excellence, which is what science and education mostly require.

The term *meritocracy* was coined by the British sociologist Michael Young in 1958, four years before Kuhn's *The Structure of Scientific Revolutions* appeared, and the proximity of these events may not be random. In a sarcastic, fairly prophetic book titled *The Rise of the Meritocracy*, Young portrays and criticizes Britain under the future rule of a class favoring merit.

Yet, what is merit? The back cover of the Pelican edition of Young's book defines it as "intelligence + effort"—an only apparently clear-cut definition, for there are multiple and distinct criteria of merit (and also of intelligence and effort), at times subjective. In fact, the book implies diverse qualities, such as competence, credentials, commitment, popularity—all steering towards what we generally call "success"—assessable in relation to a given framework. In perusing *The Structure of Scientific Revolutions*, it is amazing how one finds all these terms in reference to normal science. Education, too, although requiring excellence, above all incentivizes merit.

No doubt as Young argues, meritocracy is better equipped than past social structures, such as those based on caste, gender, race, or, more recently, membership in a political party, in achieving certain goals. Meritocracy renders a state apparatus more efficient. It helps in winning a war. It has kept empires going. The ancient Chinese rulers, inspired by Confucius, had already understood the importance of meritocracy. Similarly, the British Empire was so successful thanks to a network of civil servants hired on the basis of competitive examinations. And Napoleon was initially successful in his campaigns thanks to a meritocracy educated in the newly founded French "Great Schools." Napoleon's imperial meritocracy breached the traditional class separation and included a curious mixture of revolutionaries, army officers, and former aristocrats, including illustrious scientists (Hahn 1971, chaps. 9 and 10; Belhoste 2003, chap. 1). Today institutions such as administrations, firms, and universities encourage meritocracy. Women have been emancipated thanks to meritocracy. In general, we can thank meritocracy for much of our well-being and feeling of security.

The leading meritocrats of yesterday and today are admired and said to excel. Here lies the confusion: excellence conflicts with merit. To excel, as the Latin root suggests, means to break a framework, not to make the best of it. Meritocrats can be excellent individuals, yet their performances are normally judged in relation to specific frameworks and goals. As such, they conform; the excellent often do not. Young seems to have understood this and hence his reserve and sarcasm towards the current encouragement of meritocracy.

Kuhn, on the other hand, portrays science as giving more attention to conformism than excellence. He grants that excellence, or "genius" as he calls it, is required to "shift the vision," that is, to create a new paradigm, incommensurable with the former one (*SSR*, 115–122; this argument, by the way, is self-contradictory for one needs nonconformity to break a framework). He allows scientific leaders to be sufficiently nonconformist so as to break the framework occasionally, but normal scientists conform both in following the paradigm and in switching allegiance to a new one when told to (*SSR*, 152–153). He considers normal scientific activity, rather than excellence, the main avenue to the "*gestalt* switch" (*SSR*, 166). A new paradigm wins consensus within the scientific community that endorses the choice, irrationally or through common sense. In Kuhn's words (Kuhn 1999, 21), a scientist is successful when his endeavor "is rewarded through recognition by other members of his professional group and by them alone." Kuhn rightly produces Galileo as an example of excellence (*SSR*, 119). Yet the secret of Galileo's success was precisely in his nonconformism and ability to break scientific, literary, social, institutional, and even other boundaries. No wonder he was punished by the gatekeepers of tradition and conformism.

Admittedly, research today—both "hard" and "soft"—closely follows Kuhn's meritocratic picture. Editors of scientific periodicals and books—the springboard for academic success—as well as their purposely chosen peer reviewers, employ meritocratic criteria (Agassi 1990a). Normal scientists become opinion leaders in the wake of their popularity, rather than their contributions to science. Opinion leaders are mentioned in scholarly meetings and publications as a matter of ritual, their arguments being repeated, hailed, and embellished by trendy expressions. Universities, especially those called "centres of excellence," select and praise faculty and students according to pre-established parameters (Readings 1996, chap. 2)—an evident contradiction causing confusion. Criteria for the evaluation of projects or exam questions are formulated accordingly. Students are said to "excel" when they manage to produce a flawless, up-to-date compliance with currently accepted views. The damage is vividly described by Karl Popper in his *The Open Society and Its Enemies* (*OS*, 1, 135): instead of studying for the sake of studying, students acquire only such knowledge that is serviceable in getting over the hurdles that they must clear for the sake of their career advancement. Popper's follower Joseph Agassi labels, in Kuhnian terms, "super-normal" students those who are on the way to becoming normal scientists or academics.[1] And, for Popper, the resulting normal scientist "is a person one ought to be sorry for"; he "has been taught badly" (Popper 1999, 52).

The sticking to meritocratic criteria is easy to comprehend: a meritocratic option is safer than a violation of the framework; excellence is harder to recognize because it often takes time to be established. Yet the establishment excludes nonconformists at the risk of thereby excluding excellence as well.

Today more than ever, the need for excellence is great, inside and outside science. Kuhn and Young wrote half a century ago, in a period in which meritocracy was still needed and triumphant. Today, empires no longer die slowly and peacefully, as the British Empire did, but instead crash like the Soviet Union. That crash, incidentally, was to a great degree due to the development of science and technology, as Mikhail Gorbachev openly admitted: it came in the wake of the Chernobyl disaster (Gorbachev 2006). And many economists who caused the recent global economic crisis came from "centers of excellence." To be sure, this was and remains a complex crisis, but "centers of excellence" are considered such because they manage to create an aura of leadership and become a reference for better or worse.[2]

It is not that these centers should be closed, nor that we should encourage anarchy or altogether abandon a meritocratic approach for which we still have no substitute. Yet, it is wise to be aware of the need for excellence that differs from merit, to avoid confusing the two, and to better attend to opinions that are not quite in line with received paradigm. To reduce risk, changes of frameworks can be controlled and made gradual.[3] In any case, when requesting merit, it is advisable to clearly specify and debate the criteria for merit.

NOTES

1. Stated repeatedly in conversations with the author.
2. Not being an economist, I am not in the position to judge the responsibilities for the recent economic crash. Let me, however, refer to Joseph Stiglitz's article (2010), charging Alan Greenspan, Robert Rubin, and Larry Summers with heavy responsibilities. From *Wikipedia* (accessed July 19, 2014) I learn that Greenspan got an M.A. degree in economics at Columbia University; Rubin graduated *summa cum laude* in economics from Harvard College, and later attended the London School of Economics and received an LL.B. from Yale Law School; Summers studied at MIT and Harvard, receiving a Ph.D. from this latter university. Moreover, in 1983, at age 28, he became one of the youngest tenured professors in Harvard's history. Stiglitz was the winner of the Nobel prize for Economics in 2001.
3. By means of what Popper calls "piecemeal [social] engineering," which he considers as "the only rational one" (*OS*, 1,157).

Bibliography

Adler, Marcus N. 1907. *The Itinerary of Benjamin of Tudela.* London: Frowde, Oxford Univ. Press.

Adorno Francesco; Zangheri Luigi, eds. 1998. *Gli statuti dell'Accademia del Disegno.* Florence: Olschki.

Agassi, Joseph. 1967 [1963]. *Towards an Historiography of Science.* The Hague: Mouton; Middletown, Conn.: Wesleyan Univ. Press.

Agassi, Joseph. 1971. *Faraday as a Natural Philosopher.* Chicago: Univ. of Chicago Press.

Agassi, Joseph. 1977. "Who discovered Boyle's Law?" *Studies in History and Philosophy of Science* 8: 189–250.

Agassi, Joseph. 1981. *Science and Society: Studies in the Sociology of Science.* Dordrecht: Reidel.

Agassi, Joseph. 1984. "Training to Survive the Hazard Called Education." *Interchange* 15: 1–14.

Agassi, Joseph. 1985. *Technology: Philosophical and Social Aspects.* Dordrecht: Reidel.

Agassi, Joseph. 1987. "The Autonomous Student." *Interchange* 18/4: 14–20.

Agassi, Joseph. 1990a. "Peer Review: A Personal Report." *Methodology and Science* 2: 171–180.

Agassi, Joseph. 1990b. *An Introduction to Philosophy. The Siblinghood of Humanity.* Delmar, NY: Caravan Books.

Agassi, Joseph. 1993. *A Philosopher's Apprentice: In Karl Popper's Workshop.* Amsterdam: Rodopi.

Agassi, Joseph. 1997. "Science Education without Pressure." In *Toward Scientific Literacy: The History and Philosophy of Science and Science Teaching Proceedings of the fourth International Conference.* Edited by L. Lenz and I. Winchester, CD-ROM HPSST.pdf, The Faculty of Education, University of Calgary, pp. 1–13.

Agassi, Joseph. 1999. "Dissertation Without Tears." In Zecha 1999, pp. 59–82.

Agassi, Joseph. 2013. *The Very Idea of Modern Science: Francis Bacon and Robert Boyle.* Dordrecht: Springer.

Agassi, Joseph; Jarvie, Ian Charles. 1987. *Rationality: The Critical View.* Dordrecht: Nijhoff.

Al-Khalili, Jim. 2010. *Pathfinders: The Golden Age of Arabic Science.* London: Penguin.

Allen, Phyllis. 1949. "Scientific Studies in the English Universities of the Seventeenth Century." *Journal of the History of Ideas* 10: 219–253.

Al-Otaibi, Moneer M.; Rashid, Hakim M. 1997. "The role of schools in Islamic society: Historical and contemporary perspectives." *The American Journal of Islamic Social Sciences* 14: 1–18.

Antal, Frederick. 1948. *Florentine Painting and its Social Background*. London: Kegan Paul, Trench, Trubner & Co.

Applebaum, Wilbur, ed. 2000. *Encyclopedia of the Scientific Revolution from Copernicus to Newton*. New York: Garland.

Armytage, Walter Harry Green. 1965. *The Rise of the Technocrats: A Social History*. London: Routledge and Kegan Paul; Toronto: Univ. of Toronto Press.

Ashby, Eric. 1963. *Technology and the Academics. An Essay on Universities and the Scientific Revolution*. London: Macmillan.

Bacon, Francis. [1620 in Latin]. *The New Organon, or True Directions Concerning the Interpretation of Nature*. http://www.constitution.org/bacon/nov_org.htm. Accessed January 5, 2014. English translation based on the standard translation of J. Spedding, R.L. Ellis, and D.D. Heath in *The Works* (Vol. 8), Boston 1863.

Bacon, Francis. 1627. *The New Atlantis*. http://www.gutenberg.org/files/2434/2434-h/2434-h.htm. Accessed March 10, 2014.

Bacon, Francis. 2002. *The Major Works*. Edited by B. Vickers. Oxford: Oxford Univ. Press.

Bagnall, Roger S. 2002. "Alexandria: Library of Dreams." *Proceedings of the American Philosophical Society* 146: 348–362.

Bailey, Richard. 2000. *Education in the Open Society: Karl Popper and Schooling*. Aldershot: Ashgate.

Baldasso, Renzo. 2010. "*Portrait of Luca Pacioli as a Disciple*: A New, Mathematical Look." *Art Bulletin* 92: 83–102.

Balty-Guesdon, Marie-Geneviève. 1992. "Le *Bayt al Hikma* de Baghdad." *Arabica* 39: 131–150.

Belhoste, Bruno. 2003. *La formation d'une technocratie: L'École polytechnique et ses élèves de la Révolution au Second Empire*. Paris: Belin.

BenAicha, Hedi. 1986. "Mosques as libraries in Islamic Civilization, 700–1400 AD." *The Journal of Library History* 21: 253–260.

Ben-David, Joseph. 1971. *The Scientist's Role in Society. A Comparative Study*. Englewood Cliffs, NJ: Prentice Hall.

Bennison, Amira K. 2011. *The Great Caliphs: The Golden Age of the 'Abbasid Empire*. London: Tauris.

Benton, Christopher P. "Some Personal Reflections on Educational Practices Found in Rabbinic Literature." http://www.docbenton.com/reflections.pdf. Accessed July 5, 2013.

Beretta, Marco; Clericuzio Antonio; Principe, Lawrence M., eds. 2009. *The Accademia del Cimento and Its European Context*. Sagamore Beach: Science History Publications.

Berkey, Jonathan. 1992. *The Transmission of Knowledge in Medieval Cairo: A Social History of Islamic Education*. Princeton, NJ: Princeton Univ. Press.

Berkson, William; Wettersten, John. 1984. *Learning From Error Karl Popper's Psychology of Learning*. La Salle, IL: Open Court.

Berti, Monica; Costa, Virgilio. 2010. *La biblioteca di Alessandria: Storia di un paradiso perduto*. Tivoli, Tored.

Biagioli, Mario. 1989. "The Social Status of Italian Mathematicians, 1450–1600." *History of Science* 27: 41–95.

Biagioli, Mario. 1993. *Galileo Courtier: The Practice of Science in the Culture of Absolutism*. Chicago: Univ. of Chicago Press.

Bianca, Concetta. 1999. *Da Bisanzio a Roma. Studi sul cardinale Bessarione*. Rome: Roma nel Rinascimento.

Blackburn, Simon. 1994. *The Oxford Dictionary of Philosophy*. Oxford: Oxford Univ. Press.

Bloom, Jonathan M. 2001. *Paper Before Print: The History and Impact of Paper in the Islamic World*. New Haven: Yale Univ. Press.

Boas Hall, Marie. 1965. "Oldenburg and the Art of Scientific Communication." *The British Journal for the History of Science* 2/8: 277–290.

Boehm, Laetitia; Raimondi Ezio, eds. 1981. *Università, accademie e società scientifiche in Italia e in Germania dal Cinquecento al Settecento*, Bologna: il Mulino.

Bolgar, R. R. 1958 reprint [1954]. *The Classical Heritage and its Beneficiaries*. Cambridge: Cambridge Univ. Press.

Bollmann, Beate. 1998. *Römische Vereinshäuser. Untersuchungen zu den Scholae der römischen Berufs-, Kult- und Augustalen-Kollegien in Italien*. Mainz: Philipp von Zabern.

Boschiero, Luciano. 2007. *Experiment and Natural Philosophy in Seventeenth-Century Tuscany: The History of the Accademia del Cimento*. Dordrecht: Springer.

Brentjes, Sonja; Morrison, Robert G. 2010. "The Sciences in Islamic Societies." In *The New Cambridge History of Islam, vol. 4: Islamic Cultures and Societies to the End of the Eighteenth Century*. Edited by R. Irwin, pp. 564–639. Cambridge: Cmabridge Univ. Press.

Brentjes, Sonja. 2002. "On the Location of the Ancient or, 'Rational' Sciences in Muslim Educational Landscapes (AH 500–1100)." *Bulletin of the Royal Institute for Inter-Faith Studies* 4: 47–71.

Brentjes, Sonja. 2008. "Patronage of the Mathematical Sciences in Islamic Societies: Structure and Rhetoric, Identities and Outcomes." In *The Oxford Handbook of the History of Mathematics*. Edited by E. Robson and J. Stedall. Oxford: Oxford Univ. Press, pp. 301–28.

Brentjes, Sonja. 2014. "Teaching the Mathematical Sciences in Islamic Societies Eighth–Seventeenth Centuries." In Karp & Schubring 2014, pp. 85–107.

Brockliss, L.W.B. 1987. *French Higher Education in the Seventeenth and Eighteenth Centuries: A Cultural History*. Oxford: Clarendon.

Brown, Harcourt. 1934. *Scientific Organizations in Seventeenth Century France (1620–1680)*. Baltimore: Williams and Wilkins.

Burke, Peter. 1999. *The Italian Renaissance. Culture and Society in Italy*. Cambridge: Polity Press.

Burke, Peter. 2000. *A Social History of Knowledge. From Gutenberg to Diderot*. Cambridge: Polity Press; Oxford: Blackwell.

Camesasca, Ettore. 1966. *Artisti in Bottega*. Milan: Feltrinelli.

Camporeale, Salvatore I. 1986. "Giovanmaria dei Tolosani O.P. 1530–1546, umanesimo, riforma e teologia controversista." *Memorie Dominicane*, Nuova Serie, 17: 145–252.

Carr, David M. 2005. *Writing on the Tablet of the Heart: Origins of Scripture and Literature*. Oxford: Oxford Univ. Press.

Cascio Pratilli, Giovanni. 1975. *L'Università e il Principe. Gli Studi di Siena e di Pisa tra Rinascimento e controriforma*. Florence: Olschki.

Cecchini, Michela. 2002. *La matematica alla Corte Sabauda 1567–1624*. Turin: CRISIS.

Cennini, Cennino. 2009 [14th century] *Il libro dell'arte*. Edited by F. Frezzato. Vicenza: Neri Pozza.

Chambers, David S. 1995. "The Earlier 'Academies' in Italy." In D. S. Chambers and F. Quiviger 1995, pp. 1–14.

Chambers, David S.; Quiviger, Francois, eds. 1995. *Italian Academies of the Sixteenth Century*. London: The Warburg Institute.

Chastel, André. 1959. *Art et Humanisme à Florence au temps de Laurent le Magnifique*. Vol. 4. Paris: Presses Universitaires de France.

Cherniss, Harold. 1945. *The Riddle of the Early Academy*. Berkeley: Univ. of California Press.

Christianson, John Robert. 2000. *On Tycho's Island: Tycho Brahe and his Assistants, 1570–1601*. Cambridge: Cambridge Univ. Press.

Ciocchi, Argante. 2003. *Luca Pacioli e la matematizzazione del sapere nel Rinascimento*. Bari: Cacucci.

Clark, Peter. 2000. *British Clubs and Societies, 1580–1800. The Origins of an Associational World*. Oxford: Oxford Univ. Press.

Clark, William. 1992. "On the Ironic Specimen of the Doctor of Philosophy." *Science in Context* 5: 97–137.

Clark, William. 2006. *Academic Charisma and the Origins of the Research University*. Chicago: Univ. of Chicago Press.

Clarke, M.L. 1971. *Higher Education in the Ancient World*. London: Routledge & Kegan Paul.

Clubb, Louise George. 1965. *Giambattista Della Porta, Dramatist*. Princeton, NJ: Princeton Univ. Press.

Cobban, A.B. 1975. *The Medieval Universities: Their Development and Organization*. London: Methuen.

Cochrane, Eric. 1983. "Le accademie." In Garfagnini 1983, vol. 1, pp. 3–17.

Conant, James B. 1970. *My Several Lives. Memoirs of a Social Inventor*. New York: Harper and Row.

Cooper, Lane. 1935. *Aristotle, Galileo, and the Tower of Pisa*. Ithaca: Cornell Univ. Press.

Courtenay, William J.; Miethke, Jürgen. 2000. *Universities and Schooling in Medieval Society*. Leiden: Brill.

Crombie, Alistair C. 1969 [1952, 1959]. *Augustine to Galileo*. 2 vols. Harmondsworth: Penguin.

Crosland, Maurice. 1992. *Science under Control: The French Academy of Sciences 1795–1914*. Cambridge: Cambridge Univ. Press.

Dawson, Christopher. 2010 [1961]. *The Crisis of Western Education*. Washington, D.C.: Catholic Univ. of America Press.

Debus, Allen G. 1970. *Science and Education in the Seventeenth Century: The Webster–Ward Debate*. London: Macdonald; New York: American Elsevier.

De Gaetano, Armand L. 1968. "The Florentine Academy and the Advancement of Learning through the Vernacular: The Orti Rucellai and the Sacra Accademia." *Bibliothèque d'humanisme et Renaissance* 30: 19–52.

Dempsey, Charles. 1980. "Some Observations on the Education of Artists in Florence and Bologna During the Later Sixteenth Century." *The Art Bulletin* 62: 552–569.

de Robertis, Francesco M. 1971. *Storia delle corporazioni e del regime associativo nel mondo romano*. 2 vols. Bari: Adriatica.

Diels, Hermann; Kranz, Walther. 1951–52. *Die Fragmente der Vorsokratiker. Griechisch und Deutsch von Hermann Diels. Herausgegeben von Walther Kranz*. 3 vols. Hildesheim: Weidmann.

Dillon, John. 2003. *The Heirs of Plato: A Study of the Old Academy (347–274 BC)*. Oxford: Clarendon Press.

Discepoli di Galileo Galilei. 1975. *Carteggio 1642–1648*. Vol. 1, edited by P. Galluzzi and M. Torrini. Florence: Giunti Barbèra.

Dixhoorn, Arjan van; Speakman Sutch, Susie, eds. 2008. *The Reach of the Republic of Letters: Literary and Learned Societies in Late Medieval and Early Modern Europe*. 2 vols. Leiden: Brill.

Drake, Stillman. 1966. "The Accademia dei Lincei." *Science* 151: 1194–1200.

Drazin, Nathan. 1979 reprint [1940]. *History of Jewish Education from 515 B.C.E. to 220 C.E.* New York: Arno Press.

Duhem, Pierre. 1954. *The Aim and Structure of Physical Theory*. Princeton: Princeton Univ. Press.

Durkheim, Emile. 1965. *The Elementary Forms of the Religious Life*. New York: Free Press.

Durkheim, Emile. 1982. *The Rules of Sociological Method*. New York: Free Press.
Durkheim, Emile. 1984. *The Division of Labor in Society*. New York: Free Press.
Eamon, William. 1991 [1990]. "From the Secrets of Nature to Public Knowledge." In Lindberg & Westman 1991, pp. 333–365.
Eamon William; Paheau Françoise. 1984. "The Accademia Segreta of Girolamo Ruscelli: A Sixteenth-Century Italian Scientific Society." *Isis* 75: 327–342.
Einstein, Albert. 1905. "Zur Elektrodynamik bewegter Körper." *Annalen der Physik*, 17: 891–921.
Ephrat, Daphna. 2000. *A Learned Society in a Period of Transition: The Sunni "Ulama" of Eleventh-Century Baghdad*. Albany: SUNY Press.
Ephrat, Daphna; Elman, Yaakov. 2000. "Orality and the Institutionalization of Tradition: The Growth of the Geonic Yeshiva and the Islamic Madrasa." In *Transmitting Jewish Traditions: Orality, Textuality, and Cultural diffusion*. Edited by Y. Elman and I. Gershoni. New Haven: Yale Univ. Press, pp. 107–137.
Erskine, Andrew. 1995. "Culture and Power in Ptolemaic Egypt: The Museum and the Library of Alexandria." *Greece & Rome*, 2nd Ser., 42/1: 38–48.
Etkes, Immanuel, ed. 2006. *Yeshivot and Battei Midrash*. Jerusalem: Zalman Shazar Center & Ben-Zion Dinur Center, Hebrew University (in Hebrew).
Fantoli, Annibale. 2003a, 3rd ed. *Galileo. For Copernicanism and For the Church*. Vatican City: Vatican Observatory.
Fantoli, Annibale. 2003b. *Il Caso Galileo. Dalla condanna alla "riabilitazione." Una questione chiusa?* Milan: Rizzoli.
Favaro, Antonio. 1966 reprint [1883]. *Galileo Galilei e lo studio di Padova*. 2 vols. Florence: Successori Le Monnier.
Favaro, Antonio. 1983. *Amici e corrispondenti di Galileo*. 3 vols. Florence: Salimbeni.
Feingold, Mordechai. 1984. *The Mathematicians' Apprenticeship: Science, Universities and Society in England, 1560–1640*. Cambridge: Cambridge Univ. Press.
Feingold, Mordechai; Navarro-Brotons, Victor, eds. 2006. *Universities and Science in the Early Modern Period*. Dordrecht: Springer.
Féret, Pierre. 1913. "University of Paris." *Catholic Encyclopedia*, vol. 11. http:// en.wikisource.org/wiki/Catholic_Encyclopedia_(1913)/University_of_Paris. Accessed May 2, 2014.
Ferry, Luc; Pesron J.-P.; Renaut, Alain, eds. 1979. *Philosophies de l'université. Textes de Schelling, Fichte, Schleiermacher, Humboldt, Hegel*. Paris: Payot.
Feynman, Richard P. 1985. *Surely You're Joking, Mr. Feynman: Adventures of a Curious Character*. New York: Norton.
Field, Arthur. 1988. *The Origins of the Platonic Academy of Florence*. Princeton, NJ: Princeton Univ. Press.
Findlen, Paula. 1993. "Controlling the Experiment: Rhetoric, Court Patronage and Experimental Method of Francesco Redi." *History of Science* 31: 35–64.
Finocchiaro, Maurice A. 1989. *The Galileo Affair. A Documentary History*. Berkeley: Univ. of California Press.
Forster, Kurt W. 1971. "Metaphors of Rule. Political Ideology and History in the Portraits of Cosimo I de' Medici." *Mitteilungen des Kunsthistorischen Institutes in Florenz* 15: 65–104.
Franci, Raffaella; Toti Rigatelli, Laura, eds. 1982. *Introduzione all'aritmetica mercantile del Medioevo e del Rinascimento*. Urbino: Quattro Venti.
Frängsmyr, Tore, ed. 1990. *Solomon's House Revisited: The Organization and Institutionalization of Science*. Canton, MA: Science History Publications.
Frank, Philipp. 1953. *Einstein: His Life and Times*. New York: Da Capo Press.
Funkenstein, Amos; Steinsaltz, Adin. 1987. *Sociology of Ignorance*. Tel Aviv: Ministry of Defence (in Hebrew).
Gafni, Isaiah. 1986. *Babylonian Jewry and its Institutions in the Period of the Talmud*. Jerusalem: The Zalman Shazar Center (in Hebrew).

Galilei, Galileo, 1610. *Sidereus nuncius*. Venice.

Galilei, Galileo. 1967, 2nd ed. [1632]. *Dialogue Concerning the Two Chief World Systems—Ptolemaic & Copernican*. Translated by S. Drake. Berkeley: Univ. of California Press.

Galilei, Galileo. 1968 reprint [1890–1909]. *Le opere, Edizione nazionale*. Edited by A. Favaro. 20 vols. Florence: Barbèra.

Galilei, Galileo. 1989, 2nd ed. [1974]. *Two New Sciences Including Centers of Gravity and Force of Percussion*. Translated by S. Drake. Toronto: Wall & Thomson.

Galluzzi, Paolo. 1980. *Il mecenatismo mediceo e le scienze*. In *Idee, istituzioni, scienza ed arti nella Firenze dei Medici*. Edited by C. Vasoli. Florence: Giunti Martello, pp. 189–215.

Galluzzi, Paolo. 1981. "L'Accademia del Cimento: 'gusti' del principe, filosofia e ideologia dell'esperimento." *Quaderni Storici*, anno 16, n. 48, fascicolo 3: 788–844.

Galluzzi, Paolo. 1990. "The Renaissance Academies." In Frängsmyr 1990, pp. 303–321.

Gamba, Enrico; Montebelli, Vico. 1987. "La matematica abachista tra ricupero della tradizione e rinnovamento scientifico." In Manno 1987, pp. 169–216.

Garfagnini, Giancarlo, ed. 1983. *Firenze e la Toscana dei Medici nell'Europa del '500*. 3 vols. Florence: Olschki.

Garin, Eugenio. 1966, 5th ed. [1949]. *Educazione umanistica in Italia*. Bari: Laterza.

Garin, Eugenio. 1976 [1957, 1966]. *L'educazione in Europa 1400/1600*. Bari: Laterza.

Gattei, Stefano. 2009. "Why, and to What Extent, May a False Hypothesis Yield the Truth." In Parusniková & Cohen 2009, pp. 47–61.

Geiger, Roger L. 2015. *The History of American Higher Education: Learning and Culture from the Founding to World War II*. Princeton: Princeton Univ. Press.

Ghiberti, Lorenzo. 1998. *I commentarii* (Biblioteca Nazionale Centrale di Firenze, II, I, 333). Florence: Giunti.

Gille, Bertrand. 1964. *Les ingénieurs de la Renaissance*. Paris: Hermann.

Gillispie, Charles C. 1980. *Science and Polity in France at the End of the Old Regime*. Princeton, NJ: Princeton Univ. Press.

Gillispie, Charles C. 1997. *Pierre-Simon Laplace, 1749–1827: A Life in Exact Sciences*. Princeton: Princeton Univ. Press.

Giusti, Enrico; Maccagni Carlo. 1994. *Luca Pacioli e la matematica del Rinascimento*. Florence: Giunti.

Gliozzi, Mario. 1950. "Sulla natura dell'" Accademia de' Secreti" di Giovan Battisa Porta." *Archives Internationales d'Histoire des Sciences*, 29: 536–541.

Goitein, Shelomo Dov. 1971. *A Mediterranean Society: The Jewish Communities of the Arab World as Portrayed in the Documents of the Cairo Geniza, Vol. II: The Community*. Berkeley: Univ. of California Press.

Gorbachev, Mikhail. 2006. "Turning Point at Chernobyl." https://www.project-syndicate.org/commentary/turning-point-at-chernobyl. Accessed July 18, 2014.

Gregori, Mina; Paolucci, Antonio; Luchinat Acidini, Cristina. 1992. *Maestri e Botteghe: Pittura a Firenze alla Fine del Quattrocento*. Florence: Silvana.

Grendler, Paul F. 1989. *Schooling in Renaissance Italy: Literacy and Learning, 1300–1600*. Baltimore: The Johns Hopkins Univ. Press.

Grendler, Paul F. 2001. *The Universities of the Italian Renaissance*. Baltimore: Johns Hopkins Univ. Press.

Gutas, Dimitri. 1998. *Greek Thought, Arabic Culture. The Graeco-Arabic Translation Movement in Baghdad and Early 'Abbāssid Society (2nd–4th /8th–10th centuries)*. Oxon: Routledge.

Hadot, Pierre. 1998. *Eloge de Socrate*. Paris: Allia.

Hahn, Roger. 1971. *The Anatomy of a Scientific Institution: The Paris Academy of Sciences, 1666–1803*. Berkeley: Univ. of California Press.

Hahn, Roger. 1981. "Laplace and the Vanishing Role of God in the Physical Universe." In Woolf 1981, pp. 85–95.

Hajnal, Istvàn. 1959, 2nd ed. *L'enseignement de l'écriture aux universités médiévales.* Budapest: Maison de l'Académie des Sciences de Hongrie.

Halm, Heinz. 1997. *The Fatimids and their Tradition of Learning.* London: Tauris.

Hamarneh, Sami K. 1983. *Health Sciences in Early Islam,* 2 vols. I: San Antonio: Zahra Publications.

Hankins, James. n.d. "Humanist Academies and the 'Platonic Academy of Florence'." http://nrs.harvard.edu/urn-3:HUL.InstRepos:2936369. Accessed July 29, 2013.

Hankins, James. 1991. "The Myth of the Platonic Academy of Florence." *Renaissance Quarterly* 44: 429–475.

Harrison, Peter. 2010. "Religion and the Early Royal Society." *Science & Christian Belief* 22: 3–22.

Haskins, Charles Homer. 1975. *The Rise of Universities.* Ithaca: Cornell Univ. Press.

Heath, Thomas L., ed. 1912. *The Method of Archimedes, Recently discovered by Heiberg: A Supplement of The Works of Archimedes 1897.* Cambridge: Cambridge Univ. Press.

Helbing, Mario Otto. 1989. *La filosofia di Francesco Buonamici professore di Galileo a Pisa.* Pisa: Nistri-Lischi.

Hershberg, James G. 1993. *James B. Conant: Harvard to Hiroshima and the Making of the Nuclear Age.* Stanford: Stanford Univ. Press.

Hezser, Catherine. 1997. *The Social Structure of the Rabbinic Movement in Roman Palestine.* Tübingen: Mohr Siebeck.

Hezser, Catherine. 2001. *Jewish Literacy in Roman Palestine.* Tübingen: Mohr Siebeck.

Huff, Toby E. 2003, 2nd ed. [1993]. *The Rise of Early Modern Science: Islam, China, and the West.* Cambridge: Cambridge Univ. Press.

Humboldt, Wilhelm von. 1854. *The Sphere and Duties of Government.* London: Chapman.

Hunter, Michael. 1981. *Science and Society in Restoration England.* Cambridge: Cambridge Univ. Press.

Iamblichus. 1818. *Life of Pythagoras.* Translated by Thomas Taylor. London: Watkins.

Jaeger, Werner. 1946–47. *Paideia: The Ideals of Greek Culture.* 3 vols. Oxford: Basil Blackwell.

Jaeger, Werner. 1954, 2nd ed. [1934–1947]. *Paideia. Die Formung des griechischen Menschen.* Berlin: de Gruyter.

Jarvie, Ian C. 2001. *The Republic of Science. The Emergence of Popper's Social View of Science 1935–1945.* Amsterdam: Rodopi.

Jeauneau, Édouard. 2009. *Rethinking the School of Chartres.* Toronto: University of Toronto Press.

John Paul II. 1992. "Faith Can Never Conflict with Reason. Address to the Pontifical Academy of Sciences." *L'Osservatore Romano. Weekly Edition in English,* 4 November, pp. 1–2.

Jones, Richard Foster. 1982 [1936]. *Ancients and Moderns. A Study of the Rise of the Scientific Movement in Seventeenth-Century England.* New York: Dover publications.

Josephus, Flavius. *The Life of Flavius Josephus.* http://www.sacred-texts.com/jud/josephus/autobiog.htm.

Josephus, Flavius. *The Antiquities of the Jews.* http://www.gutenberg.org/files/2848/2848-h/book20.htm#2H_4_0001.

Josephus, Flavius. The Wars of the Jews. http://lexundria.com/j_bj/2.117-2.166/wst. Accessed November 3, 2014.

Kant, Immanuel. 1979. *The Conflict of the Faculties.* Translated by M.J. Gregor. New York: Abaris.

Karp, Alexander; Schubring, Gert, eds. 2014. *Handbook on the History of Mathematics Education.* New York: Springer.

Kepler, Johannes. 2008. *Selections from Kepler's Astronomia Nova.* Translated by W. H. Donahue. Santa Fe, NM: Green Cat Books.

Kerferd, George B. 1981. *The Sophistic Movement.* Cambridge: Cambridge Univ. press.

Keynes, Simon; Lapidge, Michael, eds. and trans. 1985 reprint [1983]. *Alfred the Great. Asser's Life of King Alfred and Other Contemporary Sources.* Harmondsworth, Middlesex: Penguin.

Kibre, Pearl; Siraisi, Nancy G. 1978. "The Institutional Setting: The Universities." In *Science in the Middle Ages.* Edited by D. C. Lindberg. Chicago: Univ. of Chicago Press, pp. 120–144.

Koestler, Arthur. 1959. *The Sleepwalkers. A History of Man's Changing Vision of the Universe.* London: Hutchinson.

Koyré, Alexandre. 1992 reprint [1968]. *Metaphysics and Measurement.* Yverdon: Gordon and Breach.

Koyré, Alexandre. 1978. *Galileo Studies.* Hassocks, Sussex: Harvester Press.

Kris, Ernst; Kurz, Otto. 1934. *Die Legende vom Künstler: Ein historischer Versuch.* Wien: Krystall Verlag.

Kris, Ernst; Kurz, Otto. 1979. *Legend, Myth and Magic in the Image of the Artist: A Historical Experiment.* New Haven: Yale Univ. Press.

Kristeller, Paul Oskar. 1945, "The School of Salerno: Its Development and Its Contribution to the History of Learning." *Bulletin of the History of Medicine* 17: 138–194.

Kuhn, Thomas S. 1996, 3rd ed. [1962]. *The Structure of Scientific Revolutions.* Chicago: Univ. of Chicago Press.

Kuhn, Thomas S. 1999. "Logic of Discovery of Psychology of Research?" In Lakatos & Musgrave 1999, pp. 1–23.

Ladous, Régis. 1994. *Des Nobel au Vatican: la fondation de l'Académie pontificale des sciences.* Paris: Cerf.

Laertius, Diogenes. 1991. *Lives of Eminent Philosophers.* Translated by R. D. Hicks. 2 vols. Reprint of the 1925 ed. Cambridge, MA: Harvard Univ. Press.

Lakatos, Imre; Musgrave Alan, eds. 1999 [1970]. *Criticism and the Growth of Knowledge.* Cambridge: Cambridge Univ. Press.

Lapidus, Ira M. 2002, 2nd ed. [1988]. *A History of Islamic Societies.* Cambridge: Cambridge Univ. Press.

Lefranc, Abel. 1970 reprint [1893]. *Histoire du Collège de France: Depuis ses origines jusqu'à la fin du premier empire.* Geneva: Slatkine.

Le Goff, Jacques. 1993. *Intellectuals in the Middle Ages.* Oxford: Blackwell.

Leiser, Gary. 1983. "Medical Education in Islamic Lands from the Seventh to the Fourteenth Century." *Journal of the History of Medicine and Allied Sciences* 38: 48–75.

Leiser, Gary. 1986. "Notes on the Madrasa in Medieval Islamic Society." *The Muslim World* 76: 16–23.

Lemaire, André. 1981. *Les écoles et la formation de la Bible dans l'ancien Israël.* Fribourg: Éditions Universitaires; Göttingen: Vandenhoeck & Ruprecht.

Le Moyne, Jean. 1972. *Le Sadducéen.* Paris: Gabalda.

Lepori, Fernando. 1980. "La scuola di Rialto dalla fondazione alla metà del Cinquecento." In *Storia della cultura veneta dal primo Quattrocento al Concilio di Trento* 3/2. Edited by. G. Arnaldi & M. Pastore Stocchi. Vicenza: Neri Pozza, pp. 539–605.

Lewis, John. 1998. *Adrien Turnebe (1515–1565): A Humanist Observed.* Geneva: Droz.

Long, Pamela O. 2001. *Openess, Secrecy, Authorship. Technical Arts and the Culture of Knowledge from Antiquity to the Reniassance.* Baltimore: The Johns Hopkins Univ. press.

Long, Pamela O. 2011. *Artisan/Practitioners and the Rise of the New Sciences, 1400–1600*. Corvallis OR: Oregon State Univ. Press.

Lindberg, David C.; Westman, Robert S., eds. 1991 reprint [1990]. *Reappraisals of the Scientific Revolution*. Cambridge: Cambridge Univ. Press.

Lucchetta, Giulio A. 1999. "Un progetto multiculturale nell'Islam dei califfi: l'ospedale." In *Saperi e scenari multiculturali, verso un'identità plurale*. Edited by E. Spedicato Iengo & G. C. Di Gaetano. Bomba (Chieti): Egidio Trailo, pp. 121–150.

Lynch, John Patrick. 1972. *Aristotle's School: A Study of a Greek Educational Institution*. Berkeley: Univ. of California Press.

Lyotard, Jean-François. 2004 reprint [1979]. *The Postmodern Condition: A Report on Knowledge*. Translated from French by G. Bennington & B. Massumi. Manchester: Manchester University Press.

Maccagni, Carlo. 1982. "Considerazioni preliminari alla lettura di Leonardo." In *Leonardo e l'età della ragione*. Edited by E. Bellone & P. Rossi. Milan: Scientia, pp. 53–67.

MacLeod, Roy, ed. 2010 reprint. *The Library of Alexandria: Center of Learning in the Ancient World*. London: Tauris.

Maffioli, S. Cesare. 1994. *Out of Galileo: The Science of Waters, 1628–1718*. Rotterdam: Erasmus Publishing.

Makdisi, George. 1981. *The Rise of Colleges: Institutions of Learning in Islam and the West*. Edinburgh: Edinburgh Univ. Press.

Malvasia, Carlo Cesare. 1969 reprint [1686]. *Le pitture di Bologna*. Bologna: Alfa.

Manetti, Antonio di Tuccio. 1970. *The Life of Brunelleschi*. University Park: The Pennsylvania State Univ. Press.

Manno, Antonio, ed. 1987. *Cultura, scienze e tecniche nella Venezia del cinquecento. Atti del convegno internazionale: Giovan Battista Benedetti e il suo tempo*. Venice: Istituto Veneto di Scienze, Lettere ed Arti.

Marrara, Danilo. 1965. *L'università di Pisa come università statale nel granducato mediceo*. Milan: Giuffré.

Marrou, Henri-Irénée. 1982. *A History of Education in Antiquity*. Madison, WI: The Univ. of Wisconsin Press.

Masood, Eshan. 2009. *Science & Islam: A History*. London: Icon Books.

Maylender, Michele. 1926–1930. *Storia delle accademie d'Italia*. 5 vols. Bologna: Forni.

McClellan, James E. III. 1985. *Science Reorganized: Scientific Societies in the Eighteenth Century*. New York: Columbia Univ. Press.

McClelland, Charles E. 1980. *State, Society, and University in Germany: 1700–1914*. Cambridge: Cambridge Univ. Press.

McLuhan, Marshall. 1962. *The Gutenberg Galaxy*. Toronto: Univ. of Toronto Press.

Mendelsohn, I. 1940. "Gilds in Babylonia and Assyria." *Journal of the American Oriental Society* 60: 68–72.

Merton, Robert K. 2001 [1938]. *Science, Technology & Society in Seventeenth-Century England*. New York: Howard Fertig.

Mettler, Suzanne. 2005. *Soldiers to Citizens. The G.I. Bill and the Making of the Greatest Generation*. Oxford: Oxford Univ. Press.

Middleton, W. E. Knowles. 1971. *The Experimenters: A Study of the Accademia del Cimento*. Baltimore: The Johns Hopkins Press.

Miert, Dirk van. 2007. "The Reformed Church and Academic Education in the Dutch Republic (1575–1686)." In *Frühneuzeitliche Bildungsgeschichte der Reformierten in konfessionsvergleichender Perspektive*. Edited by H. Schilling & S. Ehrenpreis. Berlin: Duncker & Humblot, pp. 75–96.

Moretti, Masino Glauro. 2004. *Scienza ed epistemologia in Francesco Bacone: Dal Novum Organum alla New Atlantis*. Rome: Studium.

Mueller, Ian. 1992. "Mathematical Method and Philosophical Truth." In *The Cambridge Companion to Plato*. Edited by R. Kraut. Cambridge: Cambridge Univ. Press, pp. 170–199.

Mukerji, Chandra. 2005. "Dominion, Demonstration, and Domination: Religious Doctrine, Territorial Politics, and French Plat Collection." In *Colonial Botany. Science, Commerce and Politics in the Early Modern World*. Edited by L. Schiebinger & C. Swan. Philadelphia: Univ. of Pennsylvania Press, pp. 19–33.

Mullinger, James Bass. 1911. *The University of Cambridge, Vol. III*. Cambridge: Cambridge Univ. Press.

Mullinger, James Bass; Brereton, Cloudesley; Dent, Harold Collett. 1960. "Universities." *Encyclopedia Britannica*, Vol. 22, pp. 862–874.

Musgrave, Alan. 2009. *Secular Sermons: Essays on Science and Philosophy*. Dunedin: Otago Univ. Press.

Nails, Debra. 2002. *The People of Plato. A Prosopography of Plato and Other Socratics*. Indianapolis: Hackett.

Nelson, Leonard. 1965. *Socratic Method and Critical Philosophy. Selected Essays*. New York: Dover.

Neubauer, Adolf. 1895. *Mediaeval Jewish Chronicles and Chronological Notes*. Vol. 2. Oxford: Clarendon Press. http://books2.scholarsportal.info/viewdoc.html?id=/ebooks/oca9/47/mediaevaljewishc02neub#tabview=tab1. Accessed May 15, 2013.

Nonn, Ulrich. 2012. *Mönche, Schreiber und Gelehrte: Bildung und Wissenschaft im Mittelalter*. Darmstadt: Wissenschaftliche Buchgesellschaft.

Olschki, Leonardo. 1965a reprint [1919]. *Geschichte der neusprachlichen wissenschaftlichen Literatur. Erster Band: Die Literatur der Technik und der angewandten Wissenschaften vom Mittelalter bis zur Renaissance*. Vaduz: Kraus.

Olschki, Leonardo. 1965b reprint [1922]. *Geschichte der neusprachlichen wissenschaftlichen Literatur. Zweiter Band. Bildung und Wissenschaft im Zeitalter der Renaissance in Italien*. Vaduz: Kraus.

Olschki, Leonardo. 1965c reprint [1927]. *Geschichte der neusprachlichen wissenschaftlichen Literatur. Dritter Band: Galilei und seine Zeit*. Vaduz: Kraus Reprint.

O'Malley, C.D. 1964. *Andreas Vesalius of Brussels, 1514–1564*. Berkeley: Univ. of California Press.

Oppenheimer, Aharon. 2006. "Battei Midrash in Babylon Prior to the Completion of the Mishnah." In Etkes 2006, pp. 19–29 (in Hebrew).

Ornstein, Martha. 1963 reprint from the 3rd ed. of 1938 [1913]. *The Rôle of Scientific Societies in the Seventeenth Century*. London: Archon Books.

Parsons, Edward Alexander. 1952. *The Alexandrian Library: Glory of the Hellenic World. Its Rise, Antiquities, and Destructions*. Amsterdam: Elsevier.

Parusniková, Zuzana; Cohen, Robert S., eds. 2009. *Rethinking Popper*. Dordrecht: Springer.

Pedersen, Johannes. 1984. *The Arabic Book*. Princeton: Princeton Univ. Press.

Pedersen, Olaf. 1997. *The First Universities. Studium Generale and the Origin of University Education in Europe*. Cambridge: Cambridge Univ. Press.

Peota Giuseppe. 2005. *Tra Università, collegi e accademie del Settecento italiano e francese*. Bagniaria Arsa: Edizioni Goliardiche.

Perkinson, Henry J. 1970. *Since Socrates: Studies in the History of Western Educational Thought*. New York: Longman.

Perkinson, Henry J. 1971. *The Possibilities of Error—An Approach to Education*. New York: McKay.

Perkinson, Henry J. 1982. "Education and Learning from Our Mistakes." In *In Pursuit of Truth. Essay on the Philosophy of Karl Popper on the Occasion of His 80th Birthday*. Edited by P. Levinson. Atlantic Highlands, NJ: Humanities Press, pp. 126–153.

Perkinson, Henry J. 1993. *Teachers Without Goals. Students Without Purposes*. New York: McGraw Hill.

Pevsner, Nikolaus. 1973 [1940]. *Academies of Art: Past and Present*. New York: Da Capo Press.

Pfeiffer, Rudolf. 1968. *History of Classical Scholarship from the Beginnings to the end of the Hellenistic Age.* Oxford: Clarendon Press.

Phillips, Gerald Marvin. 1956. *The Theory and Practice of Rhetoric at the Babylonian Talmudic Academies from 70 c.e. to 500 c.e. as Evidenced in the Babylonian Talmud.* Ph.D. thesis. http://www.uark.edu/depts/comminfo/sage/talmud.html. Accessed October 30, 2014.

Plaisance, Michel. 2004. *L'Accademia e il suo principe: Cultura e politica a Firenze al tempo di Cosimo I e di Francesco de' Medici; L'Académie et le Prince: culture et politique à Florence au temps de Côme Ier et de François de Médicis.* Rome: Vecchiarelli.

Plato. 1955. *Plato in Twelve Volumes,* Vol. 8. Translated by W.R.M. Lamb. Cambridge, MA: Harvard Univ. Press; London: Heinemann.

Polanyi, Michael. 1946. *Science, Faith, and Society.* London: Oxford Univ. Press. http://archive.org/stream/sciencefaithands032129mbp/sciencefaithands032129 mbp_djvu.txt. Accessed July 27, 2014.

Popper, Karl R. 1956. "Three views concerning human knowledge." In H. D. Lewis (ed.), *Contemporary British Philosophy, 3rd Series.* London: Allen and Unwin, pp. 357–388.

Popper, Karl R. 1957. *The Poverty of Historicism.* London: Routledge & Kegan Paul.

Popper, Karl R. 1965, 2nd ed. [1963]. *Conjectures and Refutations: The Growth of Scientific Knowledge.* London: Routledge & Kegan Paul.

Popper, Karl R. 1966, 5th ed. [1945]. *The Open Society and its Enemies.* 2 vols. London: Routledge & Kegan Paul.

Popper, Karl R. 1972. *Objective Knowledge: An Evolutionary Approach.* Oxford: Clarendon Press.

Popper, Karl R. 1976, revised ed. [1974]. *Unended Quest. An Intellectual Autobiography.* Glasgow: Fontana/Collins.

Popper, Karl R. 1995 reprint [1959]. *The Logic of Scientific Discovery.* London: Routledge.

Popper, Karl R. 1999. "Normal Science and Its Dangers." In Lakatos & Musgrave 1999, pp. 51–58.

Porphyry. *The Life of* Pythagoras. http://www.tertullian.org/fathers/porphyry_life_ of_pythagoras_02_text.htm. Accessed May 11, 2014.

Post, Gaines. 1967 reprint [1929]. "Alexander III, The *Licentia Docendi* and the Rise of the Universities." In *Anniversary Essays in Mediaeval History, by Students of Charles Homer Haskins.* Edited by C.H. Taylor. Freeport, NY: Books for Libraries Press, pp. 253–277.

Pryds, Darleen. 2000. "*Studia* as Royal Offices: Mediterranean Universities of Medieval Europe." In Courtenay & Miethke 2000, pp. 83–99.

Rappaport, Rhoda. 1981. "The Liberties of the Paris Academy of Sciences, 1716–1785." In Woolf 1981, pp. 225–253.

Rashdall, Hastings. 1936, 2nd ed. [1895]. *The Universities of Europe in the Middle Ages.* 3 vols. Oxford: Oxford Univ. Press.

Rashdall, Hastings. 2012 reprint [1895]. *The Universities of Europe in the Middle Ages,* vol. 2, part 1. London: Forgotten Books.

Rawson, Elizabeth. 1985. *Intellectual Life in the Late Roman Republic.* London: Duckworth.

Readings, Bill. 1996. *The University in Ruins.* Cambridge: Harvard Univ. Press.

Reale, Giovanni. 1990. *A History of Ancient Philosophy. II. Plato and Aristotle.* Albany: SUNY Press.

Redondi, Pietro. 1983. *Galileo eretico.* Turin: Einaudi.

Redondi, Pietro. 1987. *Galileo Heretic.* Princeton, NJ: Princeton Univ. Press.

Renn, Jürgen; Valleriani, Matteo. 2001. "Galileo and the Challenge of the Arsenal." *Nuncius* 16/2: 481–503.

Reynolds, Anne. 2002. "Galileo Galilei and the Satirical Poem 'Contro il portar la toga': The Literary Foundations of Science." *Nuncius* 17/1: 45–62.

Ricci, Corrado. 1888, 2nd ed. [1887]. *I primordi dello Studio di Bologna*. Bologna: Romagnoli Dall'acqua.

Riedweg, Christoph 2008. *Pythagoras: His Life, Teaching, and Influence*. Ithaca: Cornell Univ. Press.

Ross, David. 1971 reprint of the 5th ed. [1923]. *Aristotle*. London: Methuen.

Ross, Tamar. 2006. "Heed my Son, the Admonition of your Father and do not Forsake the Torah of your Mother: The History and Import of Women's Bet Midrash Movement." In Etkes 2006, pp. 443–464 (in Hebrew).

Rossi, Paolo. 1970. *Philosophy, Technology and the Arts in the Early Modern Era, 1470–1700*. New York: Harper and Row.

Rossi, Paolo. 2002, 2nd ed. [1962]. *I filosofi e le macchine, 1400–1700*. Milan: Feltrinelli.

Rossi, Sergio. 1980. *Dalle botteghe alle accademie. Realtà sociale e teorie artistiche a firenze dal XIV al XVI secolo*. Milan: Feltrinelli.

Rubin, Patricia L. 1995. *Giorgio Vasari: Art and History*. New Haven: Yale University Press.

Rüegg Walter; de Ridder-Symoens, Hilde, eds. 1992–2011. *A History of the University in Europe*. 4 vols. Cambridge: Cambridge University Press.

Russell, Bertrand. 1975, 2nd ed. *History of Western Philosophy*. London: George Allen & Unwin.

Ryle, Gilbert. 1971. *Collected Papers. Volume 1: Critical Essays*. London: Hutchinson.

Sabra, A. I. 1987. "The Appropriation and Subsequent Naturalization of Greek Science in Medieval Islam: A Preliminary Statement." *History of Science* 25: 223–243.

Sanzo, Ubaldo. 2002. "Alle origini dell'*École Polytechnique*." *Arché* 4: 227–253.

Schiffman, Lawrence H., ed. 1998. *Texts and Traditions: A Source Reader for the Study of Second Temple and Rabbinic Judaism*. Hoboken, NJ: Ktav.

Schmitt, Charles B. 1972. "The Faculty of Arts at Pisa at the Time of Galileo." *Physis* 14: 243–272.

Schneider, Ivo. 1979. *Archimedes: Ingenieur, Naturwissenschaftler und Mathematiker*. Darmstadt: Wissenschaftliche Buchgesellschaft.

Schneider, Ivo. 1993. *Johannes Faulhaber, 1580–1635: Rechenmeister in einer Welt des Umbruchs*. Basel: Birkhäuser.

Sealy, Robert J. 1981. *The Palace Academy of Henry III*. Geneva: Droz.

Segre, Michael. 1989. "Viviani's Life of Galileo." *Isis* 80: 207–231.

Segre, Michael. 1991a. *In the Wake of Galileo*. New Brunswick: Rutgers Univ. Press.

Segre, Michael. 1991b. "Science at the Tuscan Court, 1642–1667." In *Physics, Cosmology and Astronomy, 1300–1700: Tension and Accommodation*. Edited by S. Unguru. Dordrecht: Kluwer, pp. 285–308.

Segre, Michael. 1997. "Light on the Galileo Case?" *Isis* 88: 484–504.

Segre, Michael. 1998a. "The Never-Ending Galilean Story." In *The Cambridge Companion to Galileo*. Edited by Peter Machamer. Cambridge: Cambridge University Press, pp. 388–416.

Segre, Michael. 1998b. "Le biografie scientifiche all'alba della scienza moderna." *Intersezioni* 18: 403–416.

Segre, Michael. 1999. "Galileo: a 'Rehabilitation' that has never taken place." *Endeavour* 23/1: 20–23.

Segre, Michael. 2001. "Ermeneutica ebraica, ermeneutica cattolica, tradizione e scienza." In *Scienza e Sacra Scrittura nel XVII secolo*. Edited by M. Mamiani. Naples: Vivarium, pp. 55–68.

Segre, Michael. 2002. "Popper e l'educazione." In *Karl R. Popper, 1902–2002: ripensando il razionalismo critico*. Edited by S. Gattei. *Nuova Civiltà delle Macchine* 20/2: 82–88.

Segre, Michael. 2003. "Zwischen Trient und Vatikanum II: Der Fall Galilei." *Berichte zur Wissenschaftsgeschichte* 26: 129–136.

Segre, Michael. 2004. *Accademia e società: conversazioni con Joseph Agassi*. Soveria Mannelli: Rubbettino.

Segre, Michael. 2005. "Studiare senza soffrire." *Il Monitore* 39/4: 5–7.

Segre, Michael. 2008. "Galilei und die Medici: Post-Renaissance Mäzenatentum oder Postmoderne Geschichtsschreibung?" In Wolfschmidt 2008, pp. 13–29.

Segre, Michael. 2009. "Applying Popperian Didactics." In Parusniková & Cohen 2009, pp. 389–395.

Segre, Michael. 2013. *L'università aperta e i suoi nemici: radici storiche e pensiero razionale*. Lanciano: Carabba.

Settle, Thomas B. 1971. "Ostilio Ricci, a Bridge Between Alberti and Galileo." *XII Congrès International d'Histoire des Sciences: Actes, Tome III B: Science e Philosophie, XVIIe et XVIIIe siècle*. Paris: Blanchard, pp. 121–125.

Sherman, Charles P. 1908. "Study of Law in Roman Law Schools." http://digital commons.law.yale.edu/cgi/viewcontent.cgi?article=5441&context=fss_papers. Accessed April 26, 2014.

Simmel, Georg. 1906. "The Sociology of Secrecy and of Secret Societies." *American Journal of Sociology* 11: 441–498.

Sizgorich, Thomas. n.d. "The Rise of Madrasas." http://salempress.com/store/ samples/great_events_from_history_middle_ages/great_events_from_history_ middle_ages_madrasas.htm. Accessed March 28, 2013.

Sole, Giovanni. 2004. *Il tabù delle fave: Pitagora e la ricerca del limite*. Soveria Mannelli: Rubbettino.

Sorbelli, Albano. 1940. *Storia della Università di Bologna. Vol. I: il Medioevo*. Bologna: Zanichelli.

Sprat, Thomas. 1667. *History of the Royal Society of London, for the Improving of Natural Knowledge*. London: Printed by T.R. for J. Martyn.

Steinsaltz, Adin. 1989. *The Talmud. The Steinsaltz Edition. A Reference Guide*. New York: Random House.

Steinsaltz, Adin, ed. 1996. *The Talmud, vol. 25, Tractate Sanhedrin, Part 1*. New York: Random House.

Stiglitz, Joseph E. 2010. "Capitalist Fools. Five Key Mistakes that Led us to the Collapse." In *The Great Hangover: 21 Tales of the New Recession*. Edited by G. Carter. New York: HarperCollins, pp. 145–152.

Stimson, Dorothy. 1968 reprint [1948]. *Scientists and Amateurs: A History of the Royal Society*. New York: Greenwood.

Strabo. *The Geography*. http://penelope.uchicago.edu/Thayer/E/Roman/Texts/Strabo/ home.html.

Suzanne, Bernard. "Let no one ignorant of geometry enter." http://plato-dialogues. org/faq/faq009.htm. Accessed May 12, 2014.

Tartaglia, Nicolo. 1554. *Quesiti et inventioni diverse*. Venice: Bescarini.

Taton, René, ed. 1964. *Enseignement et diffusion des sciences en France au XVIIIe siècle*. Paris: Hermann.

Thoren, Victor E. 1990. *The Lord of Uraniborg: A Biography of Tycho Brahe*. Cambridge: Cambridge Univ. Press.

Toorn, Karel van der. 2007. *Scribal Culture and the Making of the Hebrew Bible*. Cambridge, MA: Harvard Univ. Press.

Torricelli, Evangelista. 1715. *Lezioni Accademiche*. Florence: nella stamp. di S.A.R., per Jacopo Guiducci, e Santi Franchi.

Torricelli, Evangelista. 1919–1944. *Opere*. Edited by Gino Loria e Giuseppe Vassura. 4 vols. Vols. 1–3: Faenza: Montanari; Vol. 4: Faenza: Lega.

Torricelli, Evangelista. 1975. *Opere Scelte*. Edited by Lanfranco Belloni. Turin: UTET.

Tosi, Anna Maria. 1989. *Il poeta dentro le mura: Ottocento carducciano e Bolognese*. Modena: Mucchi.

Totah, Khalil A. 1926. *The Contribution of the Arabs to Education*. New York: Teachers College, Columbia University.

Trexler, Richard C. 1980. *Public Life in Renaissance Florence*. Ithaca: Cornell University Press.

Tritton, Arthur Stanley. 1957. *Materials on Muslim Education in the Middle Ages*. London: Luzac.

Turner, Dorothy Mabel. 1981 reprint [1927]. *History of Science Teaching in England*. New York: Arno Press.

Univ. of Waterloo Library. Scholarly Societies Project. http://www.scholarly-societies. org/. Accessed October 27, 2014.

Vallance, John. 2010. "Doctors in the Library: The Strange Tale of Apollonius the Bookworm and other stories." In MacLeod 2010, pp. 95–113.

Van Egmond, Warren. 1980. *Practical Mathematics in the Italian Renaissance: A Catalog of Italian Abbacus Manuscripts and Printed Books to 1600*. Supplemento agli *Annali dell'Istituto e Museo di Storia della Scienza*. Fascicolo 1, Monografia n. 4. Florence: Istituto e Museo di Storia della Scienza.

Vasari, Giorgio. *Le vite de' più eccellenti pittori scultori e architettori* di Giorgio Vasari—*Edizione Giuntina e Torrentiniana*. http://vasari.sns.it/consultazione/ Vasari/indice.html. Accessed July 28, 2013.

Vasari, Giorgio. 1550. *Le vite de' più eccellenti architetti, pittori et scultori italiani, da Cimabue insino a' tempi nostri*. Florence: Torrentino.

Vasari, Giorgio. 1568. *Le vite de' più eccellenti pittori scultori et architettori*. Florence: Giunti.

Vasari, Giorgio. 1962. *La vita di Michelangelo*, ed. Paola Barocchi. 5 vols. Milan: Ricciardi.

Vasari, Giorgio. 1966–1969. *Le vite de' più eccellenti pittori, scultori et architettori nelle redazioni del 1550 e 1568*. Text (3 vols.). Edited by R. Bettarini, comments (5 vols.) of P. Barocchi. Florence: Sansoni.

Vasari, Giorgio. 1988 reprint [1965, 1971]. *Lives of the Artists*. 2 vols. A Selection Translated by G. Bull. London: Penguin Books.

Venturi, Robert; Scott Brown, Denise; Izenour, Steven. 1977 revised ed. *Learning from Las Vegas*. Cambridge, MA: MIT Press.

Verboven, Koenraad. 2007. "The Associative Order. Status and Ethos among Roman Businessmen in Late Republic and Early Empire." *Athenaeum* 95/2, pp. 861–893.

Verger, Jacques. 1999. *Les universités au Moyen Age*. Paris: Quadrige.

Visconti, Alessandro. 1950. *La storia dell'università di Ferrara (1391–1950)*. Bologna: Zanichelli.

Viviani, Vincenzio. 1968. "Racconto istorico della vita di Galileo." In Galileo Galilei, *Le opere*, vol. 19, pp. 597–632.

Vocht, Henry de. 1951. *History of the Foundation and the Rise of the Collegium Trilingue Lovaniense, 1517–1550. Part the first: The Foundation*. Leuven Librairie Universitaire.

von Staden, Heinrich. 1989. *Herophilus: The Art of Medicine in Early Alexandria*. Cambridge: Cambridge Univ. Press.

Wackernagel, Martin. 1938. *Der Lebensraum des Künstlers in der florentinischen Renaissance: Aufgaben und Auftraggeber, Werkstatt und Kunstmarkt*. Leipzig: Seemann.

Waltzing, J.-P. 1895–1900. *Ètude historique sur les corporations professionnelles chez les Romains depuis les origines jusqu'à la chute de l'Empire d'Occident*. 4 vols. Leuven Charles Peeters.

Waźbiński, Zygmunt. 1983. "La Cappella dei Medici a l'origine dell'Accademia del Disegno." In Garfagnini 1983, vol. 1, pp. 55–69.

Waźbiński, Zygmunt. 1987. *L'Accademia Medicea del Disegno a Firenze nel Cinquecento: idea e istituzione.* 2 vols. Florence: Olschki.

Weber, Max. 1946. "Science as a Vocation." In *From Max Weber: Essays in Sociology.* Translated and edited by H.H. Gerth and C. Wright Mills. New York: Oxford University Press, pp. 129–156. http://anthropos-lab.net/wp/wp-content/uploads/2011/12/Weber-Science-as-a-Vocation.pdf. Accessed October 12, 2014.

Webster, Charles, ed. 1970. *Samuel Hartlib and the Advancement of Learning.* Cambridge: Cambridge Univ. Press.

Webster, Charles. 1975. *The Great Instauration: Science, Medicine and Reform 1626–1660.* London: Duckworth.

Weizenbaum, Joseph. 1976. *Computer Power and Human Reason: From Judgment to Calculation.* San Francisco: Freeman.

Wellhausen, Julius. 2001. *The Pharisees and the Sadducees: An Examination of Internal Jewish History.* Macon, GA: Mercer Univ. Press.

Werner, Inge. 2008. "The Heritage of the Umidi: Performative Poetry in the Early Accademia Fiorentina." In Dixhoorn & Speakman Sutch 2008, vol. 2, pp. 257–284.

West, Andrew Fleming. 1971 reprint [1892]. *Alcuin and the Rise of Christian Schools.* New York: AMS Press.

Westfall, Richard S. 1980. *Never at Rest: A Biography of Isaac Newton.* Cambridge: Cambridge Univ. Press.

Westfall, Richard S. 1985. "Science and Patronage: Galileo and the Telescope." In *Isis* 76: 11–30.

Wettersten, John. 1987a. "On the Unification of Psychology, Methodology and Pedagogy: Seltz, Popper, and Agassi." *Interchange* 18/4, pp. 1–14.

Wettersten, John. 1987b. "On Education and Education for Autonomy." *Interchange* 18/4, pp. 21–25.

Whitehead, Alfred North. 1967 [1929]. *The Aims of Education and Other Essays.* New York: The Free Press.

Williams, Ronald J. 1972. "Scribal Training in Ancient Egypt." *Journal of the American Oriental Society* 92: 214–221.

Wolfschmidt, Gudrun, ed. 2008. *Astronomisches Mäzenatentum.* Norderstedt: Books on Demand.

Wolfson, Harry Austryn. 1929. *Crescas' Critique of Aristotle. Problems of Aristotle's Physics in Jewish and Arabic Philosophy.* Cambridge, MA: Harvard Univ. Press.

Woolf, Harry, ed. 1981. *The Analytic Spirit.* Ithaca: Cornell Univ. Press.

Yates, Frances A. 1988 [1947]. *The French Academies of the Sixteenth Century.* London: Routledge.

Young, Michael. 1970 [1958]. *The Rise of the Meritocracy, 1870–2033.* Harmondsworth, Middlesex: Penguin Books.

Zaimeche, Salah. 2002. "Education in Islam—The role of the Mosque." Manchester: Foundation for Science Technology and Civilisation. http://www.themodernreligion.com/misc/edu/role-of-mosque.html. Accessed October 30, 2014.

Zecha, Gerhard, ed. 1999. *Critical Rationalism and Educational Discourse.* Amsterdam: Rodopi.

Ziebarth, Erich. 1909. *Aus dem griechischen Schulwesen.* Leipzig: Teubner.

Zilsel, Edgar. 2000. *The Social Origins of Modern Science.* Dordrecht: Kluwer.

ABBREVIATIONS

DL—Laertius, Diogenes. 1991. I have used *Lives of Eminent Philosophers*. Translated by R. D. Hicks. 2 vols. Reprint of the 1925 ed. Cambridge (Mass.): Harvard Univ. Press; and *The Lives and Opinions of Eminent Philosophers by Diogenes Laertius*. Translated by c.d. Yonge, http://classicpersuasion.org/pw/diogenes/, accessed April 27, 2015.

OG—Galilei, Galileo. 1968 reprint [1890–1909]. *Le opere*, Edizione nazionale. Edited by Antonio Favaro. 20 vols. Florence: Barbèra.

OS—Popper, Karl R. 1966. *The Open Society and Its Enemies*. 2 vol. 5th ed. London: Routledge & Kegan Paul.

SSR—Kuhn, Thomas S. 1996, 3rd ed. *The Structure of Scientific Revolutions*. Chicago: Univ. of Chicago Press.

Names Index

Places Index

Subjects Index

Novum Organum (Bacon) 149
numerus clausus 118

observation, knowledge through 31–3
  see also inductive methodology/
  knowledge
observatory/ies 64, 133, 135, 153, 161,
  169n, 175
Old Testament 24, 44, 45, 47, 48, 49
  see also Bible; Holy Writ
open society 1, 25, 151, 173
*Open Society and Its Enemies,
  The* (Popper) 1, 82, 177,
  196
opinion leaders 82, 196
oral culture 93–5
Oral Torah 44, 45, 49, 50 see also
  Mishna; *Gemara*; *Talmud*
Oxford/Oxford university 89, 90, 92–3,
  96, 112, 138, 154, 156, 160,
  167, 169, 175, 180–1

Palace School (*Schola palatina*) 75, 76
"paradox of freedom" 82
Patent Office, Bern
patronage 36–7, 39, 61–3, 65, 69–70,
  77, 99–101, 107, 134, 187–94;
  Medici 100, 187–92; post-
  Renaissance 192–4
peer reviewers 158, 196
Pharisess (*Prushim*) 49–50
Philosophical Transactions 157, 158
philosophy of science 34; Popper's
  1–2, 14, 151; Bacon's 149–52;
  Descartes' 160; Kuhn's 95,
  144–5; instrumentalism 142, 147
Piemondtese Royal Military Academy
  164
Plato's Academy 3, 19–21, 25–6, 27,
  29–30, 100–1
Platonic Academy (Renaissance) 100–1
poetic knowledge *poiesis* 33
Pontifical Academy of Sciences 135
positivism 151, 155, 174, 177
post-Humboldtian university 176, 181
Postmodern Condition, The (Lyotard)
  193
practical knowledge 33–4
pre-Socratic era 11–13, 16
priesthood, Hebrew: major *see* cohen/
  cohanim; minor *see* Levites
priestly castes/class 36, 44; *collegia*, 80
Principia (*Philosophiæ Naturalis
  Principia Mathematica*, Newton)

Principles of Philosophy (Descartes)
  160
printing press 93, 94, 133, 174
private lessons 123, 128, 138, 139,
  140, 147
professional scientists 160–1
professor/s 29, 55, 86, 91, 94, 109,
  137–40, 149, 161, 166, 167,
  174, 176–7, 179–81, 183–4, 187
  see also lecturers
Ptolemaic dynasty 36, 37, 38;
  Ptolemy I Soter 35; Ptolemy II
  Philadelphus; 35–6, 38, 49;
  Ptolemy III Euergetes
  36, 38
public lessons 123, 139, 140, 153
Pythagoras 11, 12, 14, 15

Qayrawan mosque 66
quadrivium 40, 74, 76

rabbis/rav 50–3, 55, 90, 94
rector 57, 71, 86, 91, 105, 143
religion: 33, 36, 40, 44–6, 64, 72,
  88, 107, 178; monotheistic 39;
  Islamic 61, 72; secular 152,
  155, 177; atheist 177; and Plato
  19; and literacy 45; and knowledge 91; and education/learning
  44–7, 89; and higher education/
  learning 10, 15, 19–20, 46, 49,
  61, 72, 77, 85; and university 4,
  20, 80; and science 52–3, 55, 74,
  88, 91, 109, 135, 141–2, 155–6,
  177–8; and medicine 10, 38,
  92, 111; and social stability
  45–6
religious education 40, 46, 61, 70, 75
Renaissance 4, 36–7, 43n, 45, 62, 67,
  73–4, 77, 95, 97–8, 100, 107,
  113–14, 119–20, 124–5,
  142, 166, 170, 190; learning/
  knowledge/culture/science 4,
  97, 99, 113–4, 117–18, 123;
  innovation during 97–8; spirit of
  openness in 98–9; northern 102,
  125; artists 126, 190; craftsmen
  124–5; presence of institutions
  during 102–8; private circles/
  learned societies/academies
  during 53, 99, 101, 114, 118,
  126, 185; Italian universities
  99; workshops during 107–11,
  122–4, 130; scientific academies